French DNA

FRENCH DNA

Trouble in Purgatory

PAUL RABINOW

THE UNIVERSITY OF CHICAGO PRESS
CHICAGO AND LONDON

Paul Rabinow is professor of anthropology at the University of California, Berkeley. His numerous books include *French Modern: Norms and Forms of the Social Environment* and *Making PCR: A Story of Biotechnology,* both published by the University of Chicago Press.

The University of Chicago Press, Chicago 60637
The University of Chicago Press, Ltd., London
© 1999 by The University of Chicago
All rights reserved. Published 1999

08 07 06 05 04 03 02 01 00 99 1 2 3 4 5

ISBN: 0-226-70150-6 (cloth)

Library of Congress Cataloging-in-Publication Data
Rabinow, Paul.
    French DNA : trouble in purgatory / Paul Rabinow.
       p.   cm.
    Includes bibliographical references.
    ISBN 0-226-70150-6 (alk. paper)
    1. Human genetics—Government policy—France. 2. Centre
d'étude du polymorphisme humain. 3. Millennium Pharmaceuticals,
Inc. 4. Non-insulin-dependent diabetes—Research—United States.
5. Non-insulin-dependent diabetes—Research—France. 6. Human
genome. 7. Human gene mapping—France. 8. Biotechnology
industries—United States. I. Title.
QH431.R236 1999
611'.01816—DC21   99-12162
                                         CIP

To Bel y Bast

Contents

Introduction

In Paris, during the winter and spring of 1994, what was alternately characterized as a quarrel, a dispute, a struggle, a debate, a battle, or a scandal simmered and then flared up to a white-hot intensity before dissipating, as such things tend to do in Paris, as a government commission was formed to study the matter and the summer vacations approached. Immediately at issue was a proposal to institute a formal commercial collaboration between an American start-up biotechnology company, Millennium Pharmaceuticals, Inc., and France's premier genomics laboratory, the Centre d'Etude du Polymorphisme Humain (CEPH). During the early 1990s, the CEPH, led by its dynamic scientific director, Daniel Cohen, had conceived of and implemented a highly innovative and effective strategy to map the human genome. Cohen was proud to announce in December 1993 that the CEPH had won the race to produce the first physical map of the human genome. When he crafted his victory announcement, with the substantial aid of a New York public relations firm, Cohen made special efforts not to humiliate the heavily government-subsidized American laboratories whom he had just beaten.[1]

Among the leaders of the American genome mapping effort was a Massachusetts Institute of Technology (MIT) scientist named Eric Lander, a cofounder of Millennium. Daniel Cohen

was also one of the cofounders of Millennium. Scientists from
the CEPH had been discussing joint projects with scientists from
Millennium throughout 1993. Scientists from the CEPH went to
Cambridge to visit Millennium scientists and hear of their plans.
The French government had been informed of, and approved,
the idea of a commercial collaboration between the CEPH and
Millennium. The core of the collaboration was to be a project
to discover the genetic basis of non-insulin-dependent forms of
diabetes. Diabetes is a major health problem in the affluent countries, and there is good reason to believe that insights about diabetes could well be applied to obesity. The public health implications
are important. The potential market is extravagant. In order to
identify genes that might be involved in these or other conditions,
one needs as large a pool of families as possible. An examination
of inheritance patterns of these families would facilitate the search
for so-called candidate genes. Researchers at the CEPH had assembled a respectable collection of families, some of whose members suffered from non-insulin-dependent diabetes. Such collections are valuable, because they are costly and time-consuming
to assemble. Millennium proposed collaborating with CEPH scientists on the basis of the CEPH's family material. In principle,
CEPH scientists were interested in such a collaboration because
Millennium was developing potentially rapid and powerful new
technologies to identify genes (although it was cautious about entering into this new terrain). Finally, Millennium was well funded,
and the 1990s was a period of budget cutbacks for French science.
A team from Millennium went to the CEPH in February 1994
to finalize the agreement. Things came apart; confrontation, polemic, and confusion ensued. The French government moved to
block the deal. The problem, explained the government spokesman, was that the CEPH was on the verge of giving away to the
Americans that most precious of things—something never before
named in such a manner—French DNA.[2]

   French DNA's narrative is structured around these events set
against the background of contrastive developments in the United
States (the AIDS epidemic, the biotechnology industry, the Human Genome Initiative), the great specter of a possible future.[3] In
the United States, during the latter half of the 1970s, intense debate had raged about the safety (as well as ethical and philosophi-

cal implications) of what was then referred to as "recombinant DNA." An unprecedented moratorium on research devised and shepherded by leaders of the scientific community succeeded in keeping government legislation at bay and basically allaying public fears focused on the safety issue. During the early 1980s, debate shifted to the status of scientific, commercial, and ethical relationships between university- and government-based research and the nascent biotechnology industry. By the end of the decade, in the United States, the landscape had been effectively reshaped; although debate and discussion continue, a large biotechnology industry funded by a massive infusion of venture capital and an equally significant amount of capital from large, often multinational pharmaceutical companies had become an established force.[4] Millennium was not atypical of such companies; it was staffed by prestigious scientists and physicians with affiliations with Harvard and MIT and was initially funded by venture capital. From their perspective, considering an alliance with the CEPH seemed strategically astute and perfectly ordinary. They had interpreted the actions and statements of members of the CEPH during the months of preliminary contacts and negotiations as indicating that the French situation was changing in a similar fashion. In fact, this diagnosis was premature.

Arriving at the CEPH immediately after the announcement of their mapping victory, I was faced with the question of what should be the focus of my study. My entry was not a reenactment of the traditional ethnographic arrival scene on some exotic site. I was already fluent in the language, had previously lived in France for years, had just completed a complementary study of an American biotechnology company and the invention there of a powerful molecular tool, and had immersed myself in the debates and inquiries swirling around the Human Genome Initiative and its scientific, technological, ethical, legal, social, political, cultural, theological, and no doubt other dimensions. The allocation of a percentage (3–5%) of the American genome budget to social, ethical, and legal issues made it, in the words of one of its directors, "the largest ethics project in history." I was intrigued by the extravagance of this phenomenon. There had been a lot of talk of "the book of life," "the holy grail," and the like. In the early years, conferences held outnumbered genes localized. Because the genomic science itself had been successfully cordoned off from "ethical and social" scrutiny, such scrutiny was reserved only for "consequences."

At the CEPH, I soon decided that I would not concentrate on the CEPH's past triumphs. There were two reasons for this: first, I felt that there would be historians of science who would be better trained to do the archival work; second, I felt that the "genius" of the CEPH was its ability to make the next move in a manner that brought the elements into an innovative assemblage. Hence I decided to concentrate on the four research projects (aging, cancer, AIDS, parasite genomes) that Cohen had inserted (some would say imposed) into the margins of the CEPH. Introducing an anthropologist was a sort of fifth research project. I plunged first into familiarizing myself with the current molecular technology in use at the CEPH. Then the Millennium crisis happened. I kept both the experimental sites and Millennium balls in the air for the duration of my stay. Other factors intervened such that this book has taken the shape it has; that is to say, the disruption and its suite of consequences became the focus of my study.[5]

French DNA's focus is on a singular instance of a multidimensional crisis in 1994. The elements of that crisis included the felt need to transform an extremely successful and innovative large-scale scientific and technological apparatus in the face of international competition; pressing claims that the work done at the lab (and its associated allies) was of the utmost consequence not only for the future of French science but for the future well-being of humanity; acute concerns, widespread in the French cultural/political milieu, over the legitimate range of experimentation in the biosciences expressed in a vocabulary of bioethics; ferocious conflict over potential means of financing the work; personalized confrontations between leading scientists over what it meant to be a scientist today, pitting against each other (at a more general level) contrastive modes of subjectivation of science as a vocation.

French DNA is about a heterogeneous zone where genomics, bioethics, patients groups, venture capital, nations, and the state meet. Such a common place, a practiced site, eruptive and changing yet strangely slack, is filled with talk of good and evil, illness and health, spirit and flesh. It is full of diverse machines and bodies, parts and wholes, exchanges and relays. For those mortally ill, or told they are so, all this discourse, all these diverse things, can produce a good deal of anxious waiting and solicitation. It can also produce a range of other effects and affects in the world. I became intrigued by the futures being carved out of the present. Their representations ranged from ones full of dangers to others of a potential luminosity. Today, as yesterday, partisans of both visions abound. Partisans that they are, they find their antagonists' arrogance, misplaced emphases, failures of nerve, and sheer blindness trying. Amid all the discord, however, all parties agree that the future is at stake and that there is a pressing obligation to do something about it.

POSITIONING

The core of the material on the CEPH and the Association Française contre les Myopathies (AFM) was gathered through a form of extended participant observation. I spent January through June 1994 in Paris. As the project had been in preparation for some time, my contacts extend back several more years. I had previously studied the formation of the Human Genome Initiative, especially its institutionalization at the University of California, Berkeley, under the leadership of Charles Cantor. I then studied the biotechnology industry, concentrating on Cetus Corporation (later, Roche Molecular Systems) and the invention of the polymerase chain reaction. This work resulted in two books: *Making PCR: A Story of Biotechnology* (Chicago: University of Chicago Press, 1996) and *Essays on the Anthropology of Reason* (Princeton: Princeton University Press, 1996). I also served as a member of the ethics committee at the local hospital in Berkeley for several years in the early 1990s. These facts are included to alert readers to the book's background.

I had intended to work at both the CEPH and the AFM but it soon became clear that such a plan was impractical. I was fortunate that Alain Kauffmann, a Swiss sociologist, was studying Généthon (and its links with the AFM) during the same time period. We worked together regularly, comparing notes and reflections as well as visiting each other's site. We were helped by many people but special thanks go to Christian Rebollo for facilitating the work and extending truly extraordinary amounts of time, work, encouragement, and thought to help us reach our own conclusions.

Two other groups in Paris offered their hospitality and their wisdom. (1) Michel Callon and Bruno Latour, at the Centre de Sociologie de l'Innovation, provide a passage point for all working on science and technology in France. Later on, during the period *French DNA* was being written, Michel Callon and Vololona Raberharisoa were conducting an important in-depth study of the AFM. Although there was some exchange, they had agreed to a strict confidentiality agreement with the AFM and were unable to provide access to much of their material. It is of historical and ethnographic interest that I have never encountered anyone else who has been obliged to make such a contract with either industrial or academic laboratories. A preliminary overview of their findings is found in Raberharisoa and Callon, "L'implication des malades dans les activités de recherche soutenues par l'Association Française contre les Myopathies," *Sciences, sociales, et santé* 16, no. 3 (1998). (2) At the INSERM Unité, Patrice Pinel, Jean-Paul Gaudilliere, and Ilana Lowy were gracious hosts. Their work, individually and collectively, forms the core of analyses and materials for any historical analyses of the French biomedical and technoscientific community. See Patrice Pinell, *Naissance d'une fléau: Histoire de la lutte contre le cancer en France, 1890–1940* (Paris: Meraille, 1992); Anne-Marie Moulin, *Le dernier langage de la médecine: Histoire de l'immunologie de Pasteur au SIDA* (Paris: Presses Universitaire de France, 1991); Jean-Paul Gaudilliere, "Biologie moléculaire et biologistes dans les années soixante: La naissance d'une discipline, Le cas français" (Thèse pour le doctorat d'histoire des sciences de l'Université Paris VII, 1991); Ilana Lowy, "The Impact of Medical Practice on Biomedical Research: The Case of Human Leucocyte Antigens Studies," *Minerva* 25, nos. 1–2 (1987); Ilana Lowy, "Tissue Groups and Cadaver Kidney Sharing: Sociocultural Aspects of a Medical Controversy," *Journal of Technology Assessment in Health Care* 2 (1986).

During the process of direct field work and subsequently, I chose to keep a certain distance from both the Centre and the Unité, so as to reach my own conclusions. Such a strategy certainly entails paying a scholarly price; whether it entails anthropological advantages is not for me to judge.

I

Life as We Know It

Humanism has been a manner of resolving (in terms
of morality, values, reconciliation) problems that one
cannot solve at all. Remember Marx's phrase? Humanity poses only those problems to itself that it can
resolve. I believe that one can say that humanism pre-
tends to resolve those problems that it cannot pose.

MICHEL FOUCAULT, *Dits et écrits*

Max Weber, in his classic work "Religious Rejections of the World
and Their Directions," identified modern capitalism and modern
science as vectors of the generalized corrosion of human solidarity.
"Money," Weber wrote, "is the most abstract and 'impersonal' element that exists in human life. The more the world of the modern
capitalist economy follows its own immanent laws, the less accessi-
ble it is to any imaginable relationship with a religious ethic of
brotherliness." Weber's diagnosis of the icy logic of capital is well
known; less well known is his assessment of the contribution of
science, understood broadly as *Wissenschaft,* to this process. "The
intellect, like all culture values, has created an aristocracy based
on the possession of rational culture and independent of all personal ethical qualities of man. The aristocracy of intellect is hence
an unbrotherly aristocracy." Weber's claim was not that any form
of "solidarity" (a preferable term to "brotherliness") was impossible, "only"—and it is an enormous "only"—that such solidarity
could not be justified rationally, where "rationally" means in terms

7

of economic coherence or by a principle of logic internal to scientific pursuits. Ultimate value commitments were pushed into the sphere of the *irrational* or the *subjective*. For Weber, religious beliefs, in modernity, were irrational, but religious practices and institutions themselves were often disciplined and organized according to a rational means-end calculus. There was an extreme tension between eschatological beliefs and capitalism and science. Weber points to the Catholic Church's struggles with usury as an example of his claim that "[u]ltimately, no genuine religion of salvation has overcome the tension between their religiosity and a rational economy."[1] Although the Church could not "overcome" this tension, it could maintain it, and in many domains its own internal practices embodied this tension and became its vehicle.

On the side of the subject, Weber identified only "two consistent avenues for escaping the tension between religion and the economic world in a principled and *inward* manner." There is the Puritan mode of religious virtuosity; the path of rationally routinized work in this world as a means of doing God's work and testing one's state of grace that Weber made famous in *The Protestant Ethic and the Spirit of Capitalism*. In this book we will encounter a French variant of the other mode of subjectivation (and its discontents), which Weber calls this-worldly "mysticism." Its pure form is represented by "the mystic's 'benevolence' which does not at all inquire into the man to whom and for whom it sacrifices. . . . Mysticism is a unique escape from this world in the form of an objectless devotion to anybody, not for man's sake but purely for devotion's sake or, in Baudelaire's words, for the sake of the 'the soul's sacred prostitution.'"[2] Leaving aside Baudelaire's stinging metaphor, the position of principled this-wordly benevolence today is finding articulation especially in the expanding network of bioethics, which in France is hierarchically organized under the leadership of a National Ethics Committee. The committee's articulation of principles stands in an uneasy relationship with the programs for research and therapy developed and tested by the highly original patients groups brought together in the AFM, which very ardently defends those "to whom and for whom sacrifices are made," in all their specificity.

Neither (early) Durkheim's representation of social paradise as consisting in a division of labor in society that strengthens social solidarity through moral bonds of interdependence nor (late) Weber's diagnosis of civilizational hell arising from the radical separation of value spheres has proven to be quite right or entirely wrong. Rather, especially in the zone where vectors of "demagification" (*Entzauberung,* traditionally translated as "disenchantment") encounter enduring moral ties, the results have been complex, frequently inconsistent, and less definitive than any of the great social thinkers had anticipated. Especially after World War II, issues of life and death, health and disease, pathology and normality, have become matters of state. And the state, in all of its internal and external diversity, has become increasingly a congeries of institutions devoted to war and welfare, or, as some prefer, defense and social health. Welfare, and its political rationality, have become a rather more purgatorial zone of contestation, where claims to knowledge, forms of economic organization, and social values have found variant formulations. With the coming of biotechnology, not only new subjectivities and new apparatuses but also new life-forms cross divides nineteenth-century social thinkers convinced themselves were unbridgable.

The French variant under consideration here will valorize "benevolence" as the virtue through which the glacial night of American capitalism can be warmed and the practice through which both capitalism and science can be put at the service of solidarity. A problem arises, however, when events reveal that the "selflessness" of benevolent solidarity can join with the "unbrotherliness" of modern capitalism, science, and bureaucracy in unexpected and not entirely redemptive ways. Surely both the Americans and the French whom we will encounter in this book would be shocked at Weber's dour, and no doubt overdrawn, diagnosis at the end of *The Protestant Ethic and the Spirit of Capitalism:* "For of the last stage of this cultural development, it might truly be said: 'specialists without spirit, sensualists without heart; this nullity imagines that it has attained a level of civilization never before achieved.' "[3] The question remains open, however, as to whether these French and American scientists, politicians, journalists, social scientists, and ethicists would be experiencing a shock of recognition. Regardless of their subjective reactions, it is more accurate to say that Weber was wrong in his claim that he was witnessing the last stage of cultural development. The problem remains with us.

ASSEMBLAGES: SPIRITUAL AND MATERIAL

Some help in diagnosing the present conjuncture is provided by Michel Foucault's analysis of a different problematization of the body and power relations. Famously, Foucault, in one of his most lacerating formulations, concluded that in the realm of modern practices of punishment and discipline, "the soul is the prison of the body." Foucault's claim was neither a global assessment of modernity nor a metaphor. Rather, it was a specific claim that sought to identify a technology linking entities and practices. "Rather than seeing this soul," Foucault counseled,

> as the reactivated remnants of an ideology, one would see it as the present correlative of a certain technology of power over the body. It would be wrong to say that the soul is an illusion, or an ideological effect. On the contrary, it exists, it has a reality, it is produced permanently around, on, within the body by the functioning of a power that is exercised . . . on those one supervises, trains and corrects. . . . This real, non-corporeal soul is not a substance; it is the element in which are articulated the effects of a certain type of power and the reference of a certain type of knowledge, the machinery by which the power relations give rise to a possible corpus of knowledge, and knowledge extends and reinforces the effects of this power. . . . On it have been built scientific techniques and discourses, and the moral claims of humanism.[4]

Disciplinary techniques are individualizing and normative; they are not central to this book. The soul and the laboring or resisting body are not, in my opinion, the only entities productively entangled in webs of power, knowledge, and ethics today. Foucault's nomination of the soul as a milieu [*élément*] provides a helpful tool for locating a different machinery at work today.

Partially following Foucault's work, two exiled Hungarian philosophers, Ferenc Feher and Agnes Heller, propose a concept that helps to situate (if not to fully specify) the French developments under analysis. That concept is "the spiritual." It points to a new milieu beyond the soul as well as a new technology beyond discipline. The spiritual for Feher and Heller has several characteristics: (1) It has a broad interpersonal connotation that contrasts with the always individualizing Christian soul and its bodily persona. (2) It is diffuse. It applies to those actions not belonging to the "material" production of social life and serves as a collective name for everything "that is not natural." (3) The domain of the spiritual comprises that which is "not real" but which nevertheless is essential for there to be a human reality (i.e., ideas, figures of imagination, utopias, and the like). (4) To the degree that the spiritual lives in the person, it forms the "general part" of the particular (Kant's sacred humanity residing in every rational being).[5] One might say that the spiritual—interpersonal, diffuse, anthropologically distinctive, simultaneously general and particular—is an imaginary construction with very real effects and potent affects.

Feher and Heller's claim is that since moderns no longer believe in the Christian soul (*pneuma*), they must consequently also adopt a different understanding of the body. "For modernity neither the sinfulness of the vessel carrying the *pneuma* nor the message (announced by the messenger) that redemption would come from the outside were acceptable."[6] Stated in this form, their claim is too categorical, too "epochal" in the sense that one cultural world has totally replaced another one. Such a view of change has no place for, or sense of, the "contemporary" and its politics of assemblage. Weber's sociology of religion and modernity was more accurate in holding that eschatological belief and its associated practices have not disappeared under modern conditions but only been decentered. Secular moderns are but one camp. Thus, while Feher and Heller understand the "spiritual" as modernity's replacement, in a linear sense, of the body/soul couplet, I argue that what we are confronted with is less a matter of "belief" or epochal change than of an altered set of elements (some new, some old) and their configuration in practice. Today, in the domain of bioethics, what is taken to be at issue is the fate of humanity, not only in the material sense (as some environmentalists see it) but precisely in a spiritual one. What is at issue is a crisis of "dignity," the symbol enshrined in the Universal Declaration of Human Rights as the bulwark against the justification of any future Auschwitz.

What is at issue today is neither docile bodies nor sinful souls (which is not to say that such states have disappeared). What is at issue are assemblages in the process of re-formation, ones that seek to bring together health and identity, wealth and sovereignty, knowledge and values. Although the attempt to re-form the biological and social worlds with spirituality is striking, we are not witnessing a remagification of the world in Weber's sense. Weber's term *zauber* can be translated as "enchanted," as in Mozart's *Magic Flute,* but Weber employed the term in its meaning of "magic," an understanding of things in which mysterious causes are fundamentally inaccessible to scientific explanations—that is, what modern science (and capitalism and bureaucracy) had dislodged from the reigning norms of rationality. Rather, what had been (relatively) stabilized in the period following World War II in Western countries as the body, society, and ethics—and their relations—are today, again, being remade, and the assemblages in which they functioned, disaggregated. Consequently, we are witnessing, and engaged in, contestations over how technologies of (social and bodily) recombination are to be aligned with technologies of signification. The questions then are: What forms are emerging? What practices are embedding and embodying them? What shape are the political struggles taking? What space of ethics is present?

It is in the muffled movement and experimentation in and around these poorly articulated issues, rather than in the noisier scraps about beliefs and values, that a future is taking shape. Values and opinions proliferate as a matter of course in democratic, consumer capitalist societies. This proliferation does not mean that values are unimportant, quite the contrary. It means only that the value conflicts take place once people have established positions.[7] While values are plentiful, forms are rare. The invention of forms is an event. Events have a different temporality, inflect discursive and nondiscursive practices, and make new assemblages visible and describable. Then value conflicts, opinions, and taste take over. It is incontestable that major changes have been taking place in understanding, manipulating, representing, and intervening in life-forms and forms of life in both the scientific and the social senses. Technological change is an important part of the story but only a part. In the United States, for example, in the last two decades, while the most passionate value conflicts have raged around abortion, a general reshaping of the sites of production of knowledge has been occurring. To cite the biotechnology industry, the growing stock of genomic information, and the simple but versatile and potent manipulative tools (exemplified by the polymerase chain reaction) is to name a few key elements; a more complete list would include the reshaping of American universities, the incessant acceleration in the computer domains, and the rise of "biosociality" as a prime locus of identity—a biologicalization of identity different from the older biological categories of the West (gender, age, race) in that it is understood as inherently manipulable and re-formable.[8] Many of the new technologies have spread rapidly to other societies. The life-forms and forms of life are now beginning to be proposed to others. The gifts they bear will apparently be received with a certain circumspection.

Biopower

In broad strokes, Michel Foucault's concept of "biopower" as "what put[s] life and its mechanisms into the field of power/ knowledge" provides a general orientation for the investigation of these phenomena. However, it needs to be filled out and inflected if it is to illuminate the situation under study. Foucault offered a preliminary (if by now classic) formulation of "bio-technico-power" in his 1976 *History of Sexuality,* volume 1, a formulation he did not live to elaborate on (although his courses at the Collège de France during 1976–77 and 1977–78 did cover some of this ground in a very interesting fashion). "What might be called a society's 'threshold of modernity' has been reached," Foucault wrote, "when the life of the species is wagered on its own political strategies. For millennia, man remained what he was for Aristotle: a living animal with the additional capacity for political existence; modern man is an animal whose politics place his existence as a living being in question."[9] Foucault contrasted the sovereign's power of seizure unto death, his right to retaliation, and his staging of an extravagant theatricality of violence with a quieter and more insidious form of power relations that aimed at producing, nourishing, and administering forces that exerted a positive influence on life. Approaching the shift in power relations from a different angle, Foucault characterizes it as a change from "biohistory" to "biopower." Gradually, especially after the agricultural advances of the eighteenth century when more or less stable surpluses of nourishment for a substantial proportion of Europe's population began to be sustained, a new space was opening up and becoming occupied.

> This was nothing less than the entry of life into history, that is, the entry of phenomena peculiar to the life of the human species into the order of knowledge and power, into the sphere of political techniques. It is not a question of claiming that this was the moment when the first contact between life and history was brought about. On the contrary, the pressure exerted by the biological on the historical had remained very strong for thousands of years. . . . Western man was gradually learning what it meant to be a living species in a living world, to have a body, conditions of existence, probabilities of life, an individual and collective welfare, forces that could be modified, and a space for which they could be distributed in an optimal manner. For the first time in history, no doubt, biological existence was reflected in political existence. Power would no longer be dealing simply with legal subjects over whom the ultimate dominion was death, but with living beings, and the mastery it would be able to exercise over them would have to be applied at the level of life itself.[10]

Foucault identified the rise of a "normalizing society" as biopower's central distinguishing characteristic for modernity. Written in the wake of *Discipline and Punish,* with its attention to disciplinary technologies and the "colonization" of the law by the social sciences, the analytic force of that choice in illuminating a form of power relations that had previously not been identified was compelling. However, today this claim requires further thought precisely because we now have a more problematized understanding of the *bios* in biopower. That the new genomic knowledges will form assemblages with social and political networks is clear; precisely how changes in *bios* will interact with old and new forms of power relations is open to question, and the evolution must be observed and analyzed. A pressing challenge is to find and/or invent means of doing so.

The Italian philosopher Giorgio Agamben draws on a distinction pointed out by Hannah Arendt in *The Human Condition* and develops it in a provocative direction.[11] The ancient Greeks, he argues, had no single term to express what we mean by the term "life." Rather, they had two semantically distinct terms: *zoē* and *bios.* The former referred to the simple fact of being alive and applied to all living beings per se; whereas the latter term indicated the appropriate form given to a way of life of an individual or group. Philosophic discussion employed the term *bios,* since the status of life as brute existence was simply not a question worthy of extended ethical or political reflection. This is not to say that the Greeks were oblivious to the existence of mere life. For example, Aristotle observed that men were attached to sheer existence per se. In book 3 of the *Politics,* he writes: "And for the sake of mere life (in which there is possible some noble element so long as the evils of existence do not greatly overbalance the good) mankind meet together and maintain the political community. Men cling to life even at the cost of enduring great misfortune, seeming to find in life a natural sweetness and happiness (like a beautiful day)."[12] Aristotle, typically, explicitly restricts the care of "mere life," and the things necessary to ensure it, to the household, which he sharply differentiates from the political life of the polis—"Men are born with regard to life, but [exist] essentially with regard to the good life."[13] That quality which sets men off from other living beings is found in their moral and legal community, in that supplement of political life, and is intimately linked to language, which elevates humans above the level of animal existence. Of course, the Greeks' practice of slavery, their gender relations, and their whole relation to what we understand to be "nature" were radically different from our own. Sheer signs of life, of brute existence, that so concern us today in our ethical reflections on such issues as "brain death" or whether the fetus is a person would have been incomprehensible to the Greeks.

One can find Agamben's distinction of *bios/zoē* illuminating without adopting his diagnosis of Western history as a growing biopolitical nightmare. The problematization of bare life as such, in my view, touches on something central to our current situation. One of the central claims I will make is that the biopolitical articulation of *zoē* and *bios* that emerged after World War II—which centered on the "dignity" of "the human person" in response to programs to strengthen the race (or population)—is today becoming disaggregated. My claim is that the identification of DNA with "the human person" as a self-evident synecdochical relationship—the part literally stands for the whole—constitutes a "spiritual" identification. This identification is a humanistic one in Foucault's sense. To equate "the human person" with body parts or with DNA is to provide a solution to a problem that has not yet been adequately posed. "Life" is problematic today because new understandings and new technologies that are involved in giving it a form are producing results that escape the philosophical self-understanding provided by both the classical world and the Christian tradition. No new political or ethical vocabularies have adequately come to terms with it either. As the technoscientific elaboration continues, anxiety reigns in certain quarters of the city, resulting in a situation that I will later call "purgatorial."

My analysis points to the fact that the basic understanding and practices of "bare life" have been altered. The genome projects (human, plant, animal, microorganismic) are demonstrating a powerful approach to life's constituent matter. It is now known that DNA is universal among living beings. It is now known that DNA is extremely manipulable. One consequence among many others is that the boundaries between species need to be rethought; transgenic animals made neither by God nor by the long-term processes of evolution now exist. Another consequence is that the AFM and organizations like it certainly will do everything in their power to accelerate the invention of genetic therapy to alter genetic material in living humans. Whether (and when) such efforts will succeed—and more generally what the limits of manipulability of DNA are and (especially) whether DNA is the key to controlling life processes—remain open questions.

My claim is that the alliance between the CEPH and the AFM was a relatively successful initial attempt to bring *bios* and *zoē* together into a common "genomic" form. The matrix was the initial effort to map the human genome, with its focus on the search for "disease genes," and the patients associations' representation of *les malades* as a kind of universal subject. The alliance's leaders (and their multiple associates and homologues elsewhere) were lucid that this alliance was only one possible way to organize a common project. Furthermore, they were lucid that their partnership was to be *une liaison dangereuse* rather than a marriage for life. The intrigue, strategy, and coupling did prove to be fecund.

A PURGATORIAL FORM

One of the distinctive characteristics of the French situation is that the scientific practitioners and the patients groups operate under intense "purgatorial pressure." Let me explain. Unforeseeably, the ethnographic inquiry discovered that aspects of practices, discourses, and attitudes in current French experience were linked to certain religious domains. The French cultural milieu in which cutting-edge science and manufacturing operate has been distinguished in recent years by a profound uneasiness about the consequences of recent technological and scientific inventions and discoveries. Or at least that is what is claimed by those who are authorized to speak for and defend the public in such matters: bioethicists, journalists, philosophers, social critics, and a few highly visible "repentant" scientists. It is striking that in parts of their discourse they deploy aspects of the imagery, vocabulary, and—in an altered manner—even the conceptual concerns first articulated seven hundred years ago with the emergence of the zone of purgatory in Western Christianity.[14] In the moral and moralistic discursive productions surrounding key sectors of the life sciences, purgatorial themes and tropes retain (or, better, once again attain) a certain actuality. These concerns include a chronic sense that the future is at stake; a leitmotif among scientists, intellectuals, and sectors of the public turning on redeeming past moral errors and avoiding future ones; an awareness of an urgent need to focus on a vast zone of ambiguity and shading in judging actions and actors' conduct; a heightened sense of tension between this-worldly activities and (somehow) transcendent stakes and values; and a pressing need to define a mode of relationship to these issues.

These elements are all the more incisive as they are experienced and expressed by subjects who are (in the majority) forthrightly secular moderns. Some of these moderns, however, have come a long way down the historical path from what is taken to have been the unfettered and uncomplicated hopefulness and trust the Enlightenment (always in the singular) placed in reason. One finds an urgent and uneasy sense of hesitancy and caution over the consequences of a felt imperative to know and to put that knowledge into action. That ambivalence and ambiguity assert themselves most forcefully when that knowledge and those actions concern living beings. These fraught and unresolved concerns amplify the intensity of the counterreaction. Some of this counterreaction can be taken to be simply "countermodern" (almost no one is "postmodern" in France); the rest marks a more perplexed reevaluation not so much of modernity per se but what is taken to be modernity's trajectory and especially the relationship of scientific knowledge and society.[15] As the biosciences are neither limited by national boundaries nor cleanly separated from economic formations nor from the intricacies of the *raison d'état,* the impulse to legislate limits that accompanies this perplexity often yields a pathos arising from the gap between the discursive production of universal principles and effective action to truly inflect that imagined trajectory.

Jacques Le Goff's superlative *The Birth of Purgatory* recounts how a remarkable site emerged at the end of the twelfth century as well as what happened to that site in subsequent centuries.[16] Previously, the fate awaiting Christians was drastic and simple, either heaven or hell. Literally, with purgatory there appeared a middle space of waiting, of one last chance, a space to gather together all those neither entirely good nor entirely bad—surely a large portion of the flock. Of prime pertinence, this transition zone was not closed to worldly interventions; prayers of friends and relatives might well prove salvational. Thus, the universal Church articulated and accepted a new relationship between this world and what lay beyond as well as a more nuanced set of judgments

about its flock. A new pastoral machinery began to attain a pertinence and power through a new attention to the finer grains of
worldly practice articulated around the construction and promise
of a more nuanced fate. Much more than social control, the fate
of Christian souls was at stake. Significantly, human intervention
played a role in determining that fate.

The particular purgatorial space at issue here is in France.
Another distinguished historian (of death and the Revolution),
Michel Vovelle, in his *Les âmes au purgatoire,* reports that in 1994,
just a shade over seven hundred years after its appearance, 35%
of the French people believed in purgatory, and 71% of practicing
Catholics did. The same survey reported that 39% of the French
believe in extraterrestrials, 71% in mental telepathy, and 37% that
the spirits of the dead communicate with the living.[17] The Catholic
Church was a good deal more circumspect, one might say modern,
in such matters. Having spent centuries Christianizing (i.e., attempting to domesticate) popular belief and rites, and then more
recently (as Church allegiances declined) seeking ways of keeping
the flock within the fold through a strategy of a more ecumenical
tolerance of "folk" practices, the Church finds itself once again
expressing a certain caution concerning the content of popular
belief. Recent Church proclamations have prudently insisted on
limits. The French Church defined purgatory as a preliminary
"purification step that might be necessary," performed before
death, on earth. Vovelle observes, "It is a question of neither a
place nor a time; one can rather speak of a state."[18] For the
Church, then, purgatory is a state, a state of consciousness, an
element in a this-wordly spiritual technology of purification.
Vovelle's statistics do reveal a possible audience, a certain receptiv-
ity, for a discourse of fate and intercession as well as for certain
types of justificatory, condemnatory, and redemptive claims put
forward by members of the scientific and medical communities
in France. They also point to a possible demand for a type of
intervention on such issues. However, *French DNA* is not a book
about the French Church and its flock. Rather, it is concerned
with the form that intervention takes as it inflects and is inflected
by the practitioners and spokespeople, both French and non-
French, of the ethics of life, especially those invented around the
genome project.

This contemporary purgatorial space, and the practices that fill it, constitute a formative dimension of the milieu in which the events related in this book unfolded. Against this background, a more specific problem was taken up and given prominence: how best to bring capital, morality, and knowledge into a productive and ethical relationship? This issue evoked long-standing debates formerly framed in the doctrine of usury, especially the judgment of those who engaged in its practice and those authorized to make such judgments. The issue of usury had attained a renewed poignancy in the twelfth and thirteenth centuries when both the monastic orders and the larger, nonascetic population began to prosper materially. The growth of monetary relations and the increased availability, distribution, and circulation of goods of all sorts led to a revision, or at least a significant modulation, of Church doctrine concerning usury. The ever more entwined relations between moral doctrine and what would one day be capitalism unfolded gradually over subsequent centuries. Today, the always troubled relations of ethics and ethos within capitalist cultures again takes on an especially prickly tone when those authorized to speak the truth require vast sums of money to practice their sciences and thereby to produce those truths on which we so firmly believe ourselves to be dependent. The long history of counterinstitutions, whose practices of self-purification have warranted their access to verities, from the medieval monastery to the late-twentieth-century university, is once again taking a new turn. That turn is leading into a space as yet poorly known, precisely because it is very much at issue today, and because the contests over its invention remain vivid.

Another of Le Goff's books is titled *Your Money or Your Life: Economy and Religion in the Middle Ages.* It treats the changing place of usury and usurers in a postmillennium France. The book's title is drawn not from the injunctive threat uttered by a thief to his prey but from the choice faced by the usurer at the end of his life. Although few at that point would readily choose their coins over the prospect of eternal damnation, many in this increasingly challenging time were less than saintly. Church officials knew that Christian norms had not spread very far into the practices of either the peasantry or their warrior leaders. Consequently, little was demanded of the bulk of the population. Gradually, with the establishment of feudalism, the Church began to play a more engaged and attentive role in social affairs, adopting the methods of the powerful, the carrot and the stick. The stick was Satan; the carrot was purgatory; the game was fate. Eventually, the Church moderated and modulated its condemnation of usury; for example, scholastic theology introduced five acceptable excuses for usury. One unwitting step on the road to a modern capitalism was taken as usurers succeeded in finding ways to keep both their money and their (eternal) lives. Another step was the growing rationalization of doctrine, to use Max Weber's term, surrounding conditions for entering into, remaining within, and passing from purgatory.

A related set of concerns is attached to money in France: the perceived danger that monetary relations (and instrumental ones more generally) hold for moral bonds consistently identified in French thought as preeminently social. For example, Benjamin Nelson, in his book *The Idea of Usury: From Tribal Brotherhood to Universal Otherhood,* shows how the idea of usury, after a long decline, regained a certain pertinence after the Revolution. Peasant protests aimed at the Jews' practice of usury attracted Napoleon's interest. In 1806, he convoked an Assembly of Jewish Notables in Paris, seizing the occasion not only as a means of addressing the specific issue of usury but also as a means to hasten the emancipation of the Jews, by which he meant to further their integration into the French nation. Napoleon posed twelve questions to the leaders of the Jewish community, of which five concerned usury. He urged the Jews to repudiate the Deuteronomic justification for practicing usury: "the Jews should renounce clannishness by a proclamation that Jews and Frenchmen are brothers." The Jewish response was nuanced: "The Talmud made it clear that the prohibition of interest among brothers referred only to loans given to the needy, and not to commercial loans which entail risk of the capital. The law was designed to strengthen the bonds of fraternity among an isolated people." The Jewish leaders told Napoleon what he wanted to hear: "France is our country, the French are our brothers." The emperor was pleased. "Since Jews and Frenchmen [*sic*] are brothers," the emperor wrote on August 23, 1806, they "must prohibit usury in dealing with Frenchmen, or with the inhabitants of any countries where the Jews are allowed to enjoy civil rights. I am anxious to do all I can to prevent the rights restored to the Jewish people proving illusory—in a word, I want them to find in France a New Jerusalem."[19] Of course, neither the Jews nor anybody else found a New Jerusalem in Paris.

What they found instead was an attenuation of marked social boundaries as well as of the moral ties that thrive when such boundaries and their associated persecutions are strong. Nelson, consistent with a long line of French thinkers, diagnosed this situation as the crisis of modernity. "It is a tragedy of moral history that the expansion of the area of the moral community had ordinarily been gained through the sacrifice of the intensity of the moral bond, or . . . that all men have been becoming brothers by becoming equally others."[20] Today, much of this claim appears contestable: is it empirically true—or tragic—that strong boundaries and the bonds they produce are passing? In the face of militant nationalism or of fundamentalist religion or ethnic or racial essentialism, can one still assert with such assurance that all social ties per se are to be valued? After all we have learned about the historical restrictions on the public sphere from feminist historians, especially of France, it is hard to see how the passing of all forms of *fraternity* [*sic*] in-and-of-itself is to be regretted.[21] Surely, the presence of "otherness," especially "equal otherness," is not always a bad thing? Finally, one has an obligation to ask for a more complex response as to what has been the historical relationship of redemptive discourses to knowledge than the one traditionally given by defenders of the tradition of collective representations, moral bonds, and social solidarity as the highest values of humanity before joining them in waving the banner of civilization in yet another contest with barbarians at the frontiers.

Obviously an entirely different sort of analysis and scholarly apparatus would be required to demonstrate the historical and long-term ethnographic pertinence of purgatory and usury to the French situation. My purpose in introducing these terms is a different one. That purpose is captured in a perceptive review by Michel de Certeau of Le Goff's *Birth of Purgatory,* where he shrewdly calls the site opened up by purgatory "a practiced site" [*un lieu pratiqué*]. Such a place is heterogeneous, heteronomic, heteromorphic. It is composed of "stratified histories playing off each other; it is characterized by the ambiguity that makes this play possible; it is distended between heterogeneous programs which circumscribe it while encountering each other there."[22] De Certeau's identification of a common place where diversely stratified and partially incompatible histories temporarily and uneasily come together is illuminating. It is a common place with which we are familiar, even intimate. Accommodating conflicting or simply disparate practices requires a distended zone, consequently one suited to the powers of imagination and denial. De Certeau underlines the intensity invested in a zone that allows for continuing solidarity and hope. Finally, one might well ask: what function [*service*] does this space and its imaginary provide? In an unexpected and piercing fashion, de Certeau replies: "it is a call to order, an order from which the present is freeing itself."[23] Purgatorial pressure.

It is safe to say that DNA will form only one part of the emergent new understandings of life. As was stated at the outset of the Human Genome Initiative and then lost sight of in the hyperbolic discursive tidal wave of hope, fear, and metadiscourse, mapping the genome is a tool to further understanding, not its end point. We still have time. In that light, the remainder of *French DNA* is content to focus on one of contending variants of the *Epistemikos bios*.

2

Genomic Assemblages

In the early 1990s, two highly distinctive French organizations allied to put into place an exemplary new means of combining genomics, public health, and financing. The CEPH, led by a Nobel Prize winner, Jean Dausset, and a dynamic and visionary doctor, Daniel Cohen, joined forces with the AFM and agreed to attempt to map the human genome "in an industrial manner."[1] By so doing, they intended to locate and identify the specific genes thought to be responsible for the degenerative muscular dystrophies crippling and killing several thousand youngsters in France each year. Additionally, the goal was to provide a service to humanity by accelerating the means to discover "disease genes." There was an explicit impetus as well to do "world-class" science with a different ethos. Building on a partial independence from the French state based in part on novel means of financing (a private legacy left to Jean Dausset for scientific work and a widely publicized "Téléthon" reinvented for France by the AFM and its allies that provided enormous financial leverage), the CEPH and the AFM jointly created the Généthon, an innovative genomics center located in an industrial site in the suburbs of Paris. On any number of registers—scientific, technological, social, marketing—this collaboration must be deemed a success. It was not, however, stable. Not only was its financial base vulnerable, but from its inception its goals included a self-imposed obsolescence. To succeed was to transmute.[2]

The encounter between the CEPH and the AFM is certainly not the first encounter between a research apparatus and a patients group. Many such encounters have occurred in the twentieth century, especially since World War II.[3] The "conquest" of polio is exemplary in its mobilization of resources and scientific tools.[4] The long-standing "wars," "campaigns," "leagues," and "associations" formed around cancer are well known.[5] The coalitions of activists, scientists, and segments of the public and political communities during the (ongoing) AIDS epidemic unquestionably inflect the political style and social organization of such alliances.[6] In the case most closely related to the one under consideration here, Jerry Lewis and his March of Dimes and subsequent campaigns obviously formed a starting point for the efforts of the AFM. The current AFM, as we will see, is itself a transformation of a previously existing patients organization.[7] Hence, I am not making a claim of novelty in this regard nor is this topic the central one in this book. There is, however, something unquestionably "new" about the alliance between the AFM and the CEPH. As with the various patients groups and advocates in the United States (e.g., those organized around Huntington's chorea or cystic fibrosis), the advent of the genome projects opened up a distinctive scientific and technological moment that has been seized upon and elaborated differently.[8] What is distinctive—and "contemporary"—in this situation is not its radical newness but its assemblage of old and new elements. The daring decision made by the AFM to support the CEPH's effort to produce a physical map of the human genome constitutes a contemporary form. Historians will one day tell us how new and how old it appears to them.

THE CEPH: JEAN DAUSSET AND DANIEL COHEN

Jean Dausset: From Blood to Immunology
Jean Dausset was a specialist on blood, especially white blood cells. His work in the early 1950s was situated at the interface of biology

and medicine. By the late 1950s, Dausset was writing papers on the possibility of "leukocyte groups," hypothesizing that they might play a significant role in the outcome of transfusions. With the growing success at transplanting organs, such considerations took on an added potential therapeutic import. Leading transplant surgeons in France were becoming convinced that immunological types played a key role in the success or failure of organ (kidney) transplants. During this period, leading American transplant surgeons were consistently skeptical of this claim. Dausset's work honed in on identifying and naming white blood cell factors activated during transfusion. If there were immunological types, then matching donors and recipients would be extremely important to the patient's survival and flourishing. During the 1960s, it became clear that the initial optimism over finding a simple system (like red blood cell typing) that would match donors and recipients was, at best, premature. There did appear to be white blood cell/immunological configurations, but they seemed to be significantly more complicated than anticipated. Dausset's eventual identification of the proteins found on the surface of white blood cells, the "major histocompatibility complex," underscored this complexity. The human leukocyte antigen (HLA) system, as it was eventually named, became the meeting ground of hematology and immunology. Decades later, the HLA complex provided an opening onto the sumptuous genetic underpinnings of the body's boundary mechanisms and defenses.

During the course of the 1960s, Dausset (and others) showed that these proteins, whose function was still unknown, displayed an enormous variability from individual to individual. Although it was initially believed that they constituted a kind of biochemical mark of individuality, this view was eventually modified. This was a situation of great complexity, but it was proving to be a systematic complexity, and variations gradually were being painstakingly classified into systems and subsystems. Three lines of investigation emerged: the classification of HLA groups, the description of their hereditary transmission, and the development of transplantation rules. Dausset argued that white blood cell compatibility reflected tissular compatibility and hence it held the key to identifying tissue donors. To test this hypothesis, Dausset and others began the painstaking work of typing the vast complexity of the HLA system.[9]

The so-called HLA hypothesis was now in place. The hypothesis can be formulated as follows: the acceptance or rejection of organs depends on histocompatibility genes responsible for the existence of specific antigens found on the surface of white blood cells. These antigens are not, however, purely individual but are variants set within a complex combinatorial matrix. By the end of the 1960s this interpretation constituted a plausible working hypothesis but not more than that. It was difficult to verify because, among other problems, there was no standardization of serums and hence it was not possible to establish a normed typing system. Still, the HLA hypothesis can be seen to have given birth to the hope of finding "a simple, quantifiable, reproducible, and standardized means of predicting the chances of success of a kidney transplant."[10] As Anne-Marie Moulin and Ilana Lowy show, the HLA hypothesis entailed more than just a conceptual scheme. It depended intimately for its verification and application on an institutional infrastructure. Without a large population in which donors representing the entire range of variability could be identified (half of the kidneys came from cadavers), and without the possibility of transporting the organs from such donors quickly to (often distant) recipients, who had themselves already been typed and assigned a priority, there would be no chance for the system to operate successfully. In Europe the HLA hypothesis carried the day: a variety of national and international organ networks were established, most famously, Euro-Transplant.

On the other hand, if one was not convinced that a comprehensive classificatory understanding of the HLA system was the key to transplant success, then there would be no pressing need to put into place the necessarily wide-scale and expensive institutional measures required to make it work. In fact, leading American transplant specialists remained skeptical of vital elements of the hypothesis. During the late 1960s, as the European institutions began to function, initial results indicated better transplant success rates with HLA typing and matching. However, American surgeons demonstrated that there were other variables involved (organs from close kin were more readily accepted socially and immunologically than those from anonymous donors even of the same HLA type). HLA typing was clearly an important advance in understanding the immune system and appeared to improve rates of success in transplantation, but it was not a magic bullet guaranteeing successful transplantation. Hence, choices about which approach to implement depended on additional criteria.

Those additional criteria included social and ethical ones. Dausset was a committed believer in the ethical superiority of the French system of nonreimbursed public donation of blood, the sacred transaction between individuals and between those individuals and society. For HLA typing, however, because the full contours of the system were not yet understood, donation had to be organized in terms of families and consequently entailed much more systematic record keeping as well as donor commitment to respond to repeated demands for more blood. Dausset was innovative in using the media to identify and mobilize the citizenry for the cause. He made appeals in the written press and even extended these appeals to a television show. Eventually some one hundred families were chosen as reference families. The members of these families were making a serious commitment; they had to give blood every six months. Although basic transportation expenses were reimbursed, there was no other payment offered. Instead, Dausset and his small staff successfully and imaginatively deployed the symbols of public service and solidarity. Dausset organized a party for the families every year, bringing them together in Paris. This annual rite of civic solidarity, with its by now aging patron, support staff, and faithful, continued during the 1990s.

During the 1960s and 1970s, organ transplantation remained an extremely high-risk procedure, with significant mortality. A theoretical or practical consensus on HLA as the most important variable in matching donors and recipients did not exist. In France, this period was still one in which the symbolism of paternalistic professionalism and devoted concern, combined with the benevolence of the altruistic system of blood donation, constituted the ethical guarantees required for both patients groups and the state. There was no call for a national ethics committee or proposed moratoria or colloquia linking transplantation with eugenics or fascism. The dangers were not hypothetical ones looming in the future but present realities; nonetheless, they were tolerated as constituting an acceptable aspect of biomedical research. One might say that Dausset and his colleagues succeeded in establishing a framework with authorized slots for the citizenry, opinion makers, and medical mandarinate: experimentation was risky but in a climate of trust it was the only way for science and medicine to progress. Progress was what science and medicine were supposed to produce.

To do this kind of genetic work, the larger the families and the more data one had, the better: charting variability was a painstaking achievement. Dausset began to invent a system whereby the blood collected from his French families would be put at the disposition of the world community of HLA workers. He established HLA workshops where the international community of HLA researchers would meet annually to pool results. Dausset succeeded, with grace and charm and efficiency, in imposing the blood from his "one hundred families" as the material on which the world of HLA researchers worked. Dausset distributed the material for free, on condition that research results be pooled. Although such a system had clear scientific and technological advantages, it also required an institutionalized practice of networked centralization. Such centralization and standardization made Dausset and his families *incontournable* (unavoidable), as the French say.

During the 1960s and 1970s, France and the United States represented two poles of biomedical practice. The centralized model held the day in France. In the context of French solidarity—"all equal in transplant" [tous égaux devant la greffe]— HLA typing provided an objective criterion for selection. There was a national list for transplants, and a recipient was assigned a place on that list. A tacit premise of the European approach was that waiting time was not crucial in the success of the procedure; when this assumption was shown not to be true, the French response was to accelerate the harvesting of organs. In the United States, certain regions adopted an approach similar to the French one (e.g., in the region surrounding Boston and Cambridge, an area rich in hospitals and public health services). In other American regions, this centralized and standardized approach was not adopted, either because it was too expensive to implement or because it had not been shown to be unequivocally superior in the eyes of surgeons or insurers. There was, of course, no national system of insurance or health care. Consequently, the type of care one received in the United States varied with one's geographic and social location. The lack of a national health care system, the tradition of local autonomy, the ambiguities in the statistics, the insurance system, and, as Anne-Marie Moulin rightly points out, the greater willingness to accept inequalities in life chances based on class, race, and gender all worked toward a patchwork aggregation of different approaches in the United States.

Despite the clear scientific advances being made in typing the HLA system, both American and European statistical studies in the 1970s showed that success rates in transplantation were not reducible to HLA factors alone. Thus, while HLA typing had become a standard test in all transplant centers, not all of them, and especially some of the most successful in the United States, had made HLA the key factor for matching donors with recipients. That being said, during the decade of the 1980s, the polar contrast between the two continents became significantly less stark. Medicare began paying for transplants in the United States, thereby including a significant population that had previously been excluded for financial reasons. It also became generally recognized that multiple factors (social, lifestyle, hematological, immunological, medical, etc.) were involved in successful transplants. Debate continues today about the appropriate weighting of such factors but the debate is no longer carried on in missionary terms; total commitment to the HLA hypothesis has been tempered in Europe just as HLA typing has become standard in the United States.

Daniel Cohen: From Blood to Genes
the Industrial Way

When Daniel Cohen arrived in Dausset's lab in experimental biology at the Collège de France in 1978 to do an internship in immunogenetics before beginning an internship in a medical specialty he had not yet chosen, no direct DNA work was being done in Dausset's lab.[11] Cohen familiarized himself with the painstaking HLA typing—science progressing the way it sometimes does, by small increments. HLA studies were undertaken at the protein level. Cohen claims credit for suggesting to Dausset that the charting of polymorphisms would be infinitely easier at the DNA level, if anyone could invent a means for doing it. In retrospect, Cohen's logic is impeccable: "the infatuation for DNA comes from the extraordinary facility with which one can analyze it. . . . In reality, DNA is a dead, inert molecule, a simple matrix. Proteins, on the other hand, are a living substance, extraordinarily sensitive and reactive."[12] Cohen, like others at the time, prophesied that the future of all such work would depend on technological innovation. Dausset was responsive to Cohen's arguments and funded Cohen, Howard Cann, and Luis Ascano to work for a year to see if they could describe the DNA variability of an HLA gene. Cohen did a further internship in genetics at another hospital in Paris before returning chez Dausset.

In 1980 Dausset was awarded the Nobel Prize for medicine and physiology for his work on HLA. This distinction advanced him into an extremely small circle of Nobel Prize winners in France. In 1981, in the wake of the publicity surrounding the prize, Hélène Anavi, a rich art collector, left her collection, worth 50 million francs, to Dausset. The gift put Dausset in a unique position. Not only was he at the absolute summit of France's institutional and prestige structure, he now had an important degree of financial independence. This independence opened the possibility of basically unrestrained, rapid, and innovative action. To achieve these goals, Dausset established the CEPH in 1984.

This potential was not lost on Cohen. He reasoned that although the international HLA workshops and reference family panel were an excellent combination, the model was vulnerable. Given the unequal resources that existed between the French scientific world and the American, the artisanal approach to collection and distribution needed to be "industrialized" if Dausset was to maintain his strategic advantages and his prominence. The crucial first step in Cohen's view was to establish a better method of ensuring the supply of raw materials; the altruistic blood system was simply too limited in how much blood it could supply, and ultimately too costly and too dependent on the network of personal relations Dausset had built up over the years. Cohen urged Dausset to explore newly developed techniques for "immortalizing" white blood cells as a potential substitute for the repeated appeals for donations. By injecting a cancer-causing virus (Epstein-Barr) into white blood cells, these cells could be made to continue dividing indefinitely. Hence the problem of shortage of materials would be solved. "By so doing, you establish an inexhaustible bank of human genomes."[13] The same materials could essentially be used over and over again. One act of civic altruism—or remunerated exchange, as blood collected for medical research was regulated in a different fashion—could be turned into an indefinite future of self-replicating material.

Dausset and Cohen realized that their families would provide an invaluable scientific tool and institutional advantage in looking for not just genetic markers of the HLA system but markers of any kind. Initial work on identifying the location of genes was being done piecemeal. Such an approach made comparison and standardization difficult. Mapping the same families would greatly facilitate the task of standardization and understanding. Cohen saw that the "one hundred families" could serve not only as the standard for the HLA system but for the genome as a whole. With the new techniques to immortalize cells, it would be possible to scale up to an entirely different type of operation. Such a strategy, Cohen reasoned and Dausset listened, would be an enormous coup; it would transform the HLA model (with its slow pace, incremental progress, network of collaborators, altruistic base, and civic mission) into a DNA model that Cohen and Dausset correctly predicted as the key future direction but whose contours remained to be imagined and invented. Certain elements of the overall strategy put into place by Dausset for the HLA system could remain the same: guaranteeing the quality and uniformity of the DNA and maintaining control over distribution of the core biological materials, thereby safeguarding the laboratory's role as obligatory passage point for collecting and distributing results. "Everything was there: a disinterested and humanistic scientific initiative, an obligatory passage that could work for generations of researchers. All the elements of a scientific service with a universal vocation were united."[14] All the elements for a major institutional and scientific coup were also in place. The money from Mme Anavi would pay for the distribution, giving them plenty of room to maneuver. "From one perspective, the altruism of the system made it very competitive. This 'gratuity' was the apple of our eye and we pushed the principle as far as we could."[15] Cohen had forged the symbolic link with the older system, thereby reassuring Dausset that all his ethical principles would be respected. While Cohen respected those principles, the apple of his eye was clearly strategic.

To begin implementing their genome project, which would culminate in the formation of the CEPH, all Cohen and Dausset needed to do was to convince the Americans. They were not alone in formulating plans for genome mapping. The American geneticist Raymond White took the lead in 1979 in the first efforts to use restriction fragment length polymorphisms (RFLPs) to map the human genome. He moved to Utah to work on the extensive multigenerational Mormon pedigrees, supported by research funding from the Howard Hughes Medical Institute. The combination of a wealthy private foundation, a university-based research team, and special access to distinctive materials proved highly productive.

During 1979–81, the U.S. government's National Institutes of Health (NIH) were approached by highly reputable scientists with a project for establishing a linkage map of the human genome. The NIH considered the proposals and decided not to support them. While White and his team proceeded to work on their Mormon pedigrees and while individual teams around the country (and world) were doing linkage studies directed at identifying specific disease genes, a private biotechnology company, Collaborative Research, Inc., expressed initial interest in developing a project to produce a linkage map of the human genome. Helen Donis-Keller, a Harvard Ph.D. who had joined her advisor (Nobel Prize winner Walter Gilbert) in his start-up biotech company, Biogen, left in 1983 to join Collaborative Research. In these early days of the American biotech industry, new ventures were initiated and supported by the prestige of the cream of the biomedical research establishment, many of whom were young and ambitious for rewards both symbolic and monetary as well as eager to see their science advance as fast as possible. Heading Collaborative Research's scientific advisory board was another Nobel Prize winner, David Baltimore, then of MIT. The company applied for an NIH Small Business Innovative Research award but was told their project to produce a linkage map of the human genome was not "sufficiently innovative."[16] Having failed to raise funds in the public sphere, the company then sought funding on Wall Street, doing presentations at large pharmaceutical and chemical companies such as Johnson & Johnson and Union Carbide. Donis-Keller had developed a sophisticated strategy for mapping markers and finding genes, thereby opening up the possibility of genetic diagnostics as well as a means of identifying potential therapeutic targets. Rebuffed, "she was appalled at the lack of vision among American corporations. They simply did not believe that genetics would be critical to cancer, heart disease, or other major health problems that bred diagnostic and therapeutic markets."[17] Collaborative Research decided to underwrite the project from their own funds, and by the end of 1984 they supported twenty-four people to work on it. During the mid-1980s, therefore, gene mapping was not yet on the map, as it were; money and scientific investment went instead into gene hunting. The fierce competition to "find the gene for . . ." became a costly and highly publicized activity into which money flowed; the major scientific journals became key players, with prepublication press conferences and the like—for many, a shocking change in professional demeanor and ethics.

A fierce competition unfolded over the course of the next several years between these two hybrid competitors: Ray White with his Mormon pedigrees and Howard Hughes support, and Helen Donis-Keller and her start-up biotech base with its Nobel Prize–winning scientific advisors. The showdown came in 1987. White and his team, the rumor had it, were winning the race and would announce their victory in Paris at a scientific conference. Shortly before the event, Collaborative Research submitted the first genetic linkage map to the prestigious journal *Cell*. In this volatile period in the United States in which the boundaries between industry-, government-, university-, and foundation-based science were rapidly being drawn and, after negotiations, redrawn, criticism flared up. Donis-Keller and her team were criticized from several quarters for what was considered to be a premature rush to publication before their markers could be confirmed. Further censure was leveled at Collaborative's use of markers identified by others who had put them into the public domain. Collaborative's private status and potential patenting issues also inflamed passions. Robert Cook-Deegan, in his authoritative history of the Hu-

man Genome Initiative, writes: "The glue holding the various genetic linkage efforts together, despite the rivalries and tensions,
was the CEPH. CEPH was formed with the express purpose of
enabling groups to pool their efforts in constructing a complete
genetic linkage map. The first meeting to piece together the coali-
tion took place in November 1984, when major mapping groups
from Europe and North America descended on Paris, or rather
ascended to a hallowed terrain where common good transcended
rivalry."[18] Although some of the CEPH reference families had
been included because of specific diseases like Huntington's, the
basis of the CEPH's strategy lay in the project to develop a linkage
map of the whole genome through a systematic approach to placing markers. These markers could, of course, then serve as a guide
to identifying the location of disease genes. As Cook-Deegan observes, American "[g]overnment funding agencies appeared oblivious to this distinction between gene-hunting and systematic mapping."[19] In 1984, as we have seen, the biomedical industry was
oblivious to the distinction as well.

In 1980, Dausset went to visit Raymond White in Utah and
proposed a collaboration that White, with the advantages provided by his extensive Mormon pedigrees and Howard Hughes
funding, understandably rejected. As Dausset made moves to tool
his lab for the nucleic acid work and made a number of other
strategic alliances, White eventually changed his mind. So, too,
did Collaborative Research. The Cohen-Dausset strategy was
working. In 1984, the CEPH put together the pieces of its collaborative international research on DNA from the same families. The
core included Mormon families from Utah, samples from an extended family with Huntington's disease in Maracaibo in Venezuela, and the CEPH's reference family panel. The families, again,
were chosen for their large sib size rather than for disease characteristics.[20] By 1986, fifty labs were taking part in the collaboration.

Bernard Barataud and the AFM:
The Status Quo Is Intolerable

In 1989, the CEPH obtained the support of the European Community for the transformation of its family-based linkage studies

into a physical mapping project, titled Predictive Medicine. In April 1990, the NIH and Department of Energy (DOE) released their first five-year plan to map the human genome. In June, Hubert Curien, France's minister of science and technology, announced the government's intention to commence a genome research program. Its proposed funding was $1.4 million to be distributed by a committee headed by Dausset. Future, higher levels of funding were announced in October but were never actually disbursed. As the Americans had allocated (and would disburse) close to $200 million in 1990, it is clear that the French government did not intend to compete with the Americans. There was strong opposition in France to the genome mapping project from a variety of scientific and governmental quarters for reasons similar to those that had been put forward in the United States.

AFM Militants

There existed, however, a nongovernmental actor on the French scene whose intervention would radically change the landscape. The AFM is the successor to a number of preexisting dystrophy patients groups. These groups had been active in small-scale mutual aid, doing what they could to promote solidarity with those like themselves exposed to the same relentless decline of their children. As was typical of the older order of things, these groups were primarily clinic oriented, engaged in supporting improvements in such things as prosthetics and pharmacology. The dystrophies fell into a category of what was called in the United States "orphan diseases," which afflict a relatively small population and for which little diagnostic or therapeutic progress had been made in a long time, and for whose sufferers nothing positive could be discerned on the horizon. Patients groups of this sort tended to be weak and divided. This situation began to change radically with the entry into the grim dystrophy world of Bernard Barataud. Barataud is an employee of the EDF (state electrical service). His three-year-old son was diagnosed with Duchenne muscular dystrophy, a degenerative and ultimately fatal disease attacking the muscles (first limiting mobility and ultimately the chest muscles and hence the ability to breathe). Like Cohen, Barataud has a strong personality, which found expression in his labor union commitments and activities, as well as in what the French call *associational* (social) attachments. He also had a strong sense of social justice. Barataud was enraged by what he considered to be the general arrogance complemented by near total ignorance on the part of the medical, social security, and research institutions concerning the dystrophies and the fates of those afflicted by them. He focused that rage into organizational and mobilization efforts, ever aware that he was losing his race against the clock, monitoring his son's condition as it worsened, engaged in a spiral of palliatives and despair. Barataud's son and the family and friends endured more than their fair share of harrowing incidents of exemplary incompetence and callousness that hastened the child's untimely demise. Barataud brought suit for malpractice against the poorly trained emergency services and the doctors and hospital administrators who had refused to admit his son as his emergency helicopter circled the Parisian skies seeking a hospital that would admit the child. After a year's wait, a formal hearing was held, and the case dismissed. It was no chance event, Barataud is convinced, that the hearing was held in a morgue.

Barataud opens his book, *Au nom de nos enfants,* with the following ferocious indictment, a summation of more than a decade of relentless activity: "I did not choose my camp. It was imposed on me by fate. Because all the official powers abandoned us, we had no other choice but to choose a breakthrough. We instituted the Téléthon. But without genetic knowledge advancing it would not be worth much. So, we created the Généthon laboratory."[21] The language of enraged revolt, of military tactics and solidarities, of a passionate commitment to scientific advance as the only hope, is perhaps not atypical of the frustration endured by countless thousands of other relatives of those afflicted with major maladies. What Barataud did with those emotions, however, is unprecedented in France. The Wexler family in the United States and Barataud in France put "their" genetic condition high on what had previously been a nonexistent research agenda. The possibility of doing so obviously depended on there being a molecular agenda at all.

Barataud's metaphors have a specific French cast. In a tone of infuriated denunciation, he exposes the arrogance and indifference of those who he feels treated him this way. Barataud indicts the "shame" of the medical profession (always, however, with a few heroic exceptions). After the appearance of his son's initial symptoms, the family consulted twelve doctors. "Twelve impeccable suits with bow ties and cuff links. Twelve bored mimics. Bored but superior. Twelve different diagnoses. Of the first twelve, not one admitted his incompetence. This young boy was on the way down." Or, later on, after another consultation requiring a long trip and resulting in another quick dismissal: "Suddenly, I wanted to demolish the pink face of this Vichy collaborator of health." But, for Barataud, it is not only the medical profession that stands accused but the moral character of France itself: "My beautiful little kid . . . alive, so sweet, was dissolving, diminishing. A little like those millions of men who were dissolved into night and fog, in the midst of those who knew without knowing, who had become deaf, blind, mute, impotent." Beyond their hateful complacency and complicity, Barataud casts the fatalism as a fundamental betrayal: "Unable to provide care, uninterested in these kids, the doctors attempted to justify a strategy of programmed death. Everything was abandonment, resignation, and ignorance. This strategy is found in health care costs, numbers, ethical reasonings, as well as technical and even humanitarian ones. A few bastards evoked such terms as therapeutic excess." One sees here a refusal of the banality of evil. Again, this evil and this banality find in France a specific historical resonance. "The right word, the one that hits its target and casts aside the others, strangely enough came from a doctor who said, 'It's like killing a member of the Resistance.'"[22] There is no more powerful symbol in contemporary French culture.

Pushed by a sense of quotidian pain and loss among the circle of families knitted together in the patients groups, and by his inquisitorial fury, Barataud rose to prominence and power in the tiny myopathy world. Barataud and his allies overcame the debilitating sectarianism arising from the previous division into ineffective and rival organizations. They eventually achieved unity; this propelled fund-raising efforts and increased their credibility as they forged links with medical and scientific allies. Barataud and his inner circle, however, gradually came to understand that money, while absolutely essential, was not sufficient. After all, Jerry Lewis and his telethons had raised enormous amounts of money without fundamentally changing the situation of "Jerry's kids." For Barataud, the only path to eventual cures lay in science, in genomics. The "means" that money could buy increasingly came to signify science, and science increasingly came to mean genetics. In 1981, the AFM decided to assign a primacy to research efforts. To that end, they established a high-profile scientific council. Making such a decision was not easy, as it meant diverting money from the immediate concerns of the afflicted children and their families. It was not universally applauded within either the dystrophy community or the French genetics world: scientific research was risky and there were pressing needs already in place. After much discussion and soul-searching, the AFM forged a model they would stick to with tenacity, a strong division of labor within a clear decision-making hierarchy. The model was more military than bureaucratic or industrial. The scientific council had the independence to make its own judgments on research proposals submitted to the AFM. There was a formal delegation of trust but one that was constantly scrutinized. Because Barataud and his colleagues were not mincing words or treading lightly on reputations and strongly entrenched professional interests in France, they went to great lengths to institute a fiscal "transparency," knowing that counterattacks would come, as they certainly did.

By the mid-1980s, Barataud had become the driving force to make genetics and the search for disease genes the priority of the myopathy world. The "gene" became the key symbol, the embodiment of fate, the evil locus from which arose death and ruination of innocent life, and, simultaneously, the site of hope. This focus is captured by Barataud in the following passage describing events at the AFM's second annual research conference in 1986:

> Suddenly, a note is passed to me, unexpected and stupefying. All the AFM members please come to the auditorium in three minutes. . . . [Standing there was] a young American, Anthony Monaco, 26 years old. . . . And suddenly an absolute silence turns magical. On the wall were blue-lit slides; whether or not one understood English it was obvious that the American was announcing a discovery, the X chromosome, Duchenne muscular dystrophy; the

gene was located on XP21. *The* gene. The origin of
Alain's [his son] disease here on the screen before me. For
the first time the beast is visible.

Good God, this was no old distinguished scientist who
was announcing this to us with his Legion of Honor rib-
bon in his buttonhole. It is that adolescent in sneakers
and jeans who looks like he just left a party.

Joy. Pain. Anger.[23]

For Barataud, the discovery of the Duchenne gene was an
element in a world revolution. A revolution that France was not
participating in. The official French response was hopeless. "The
state has no strategy to conquer these diseases. One thing is sure:
there is nothing to hope for from the government. Finding a way
out is perfectly legitimate and indispensable. First comes knowledge. Without knowledge there is no hope. Each year medical
errors kill as many in our ranks as the disease itself. We must lay
the foundations for a strong, multidisciplinary research effort."[24]
Given this bleak evaluation, the AFM assigned the highest priority
to "communication" and public fund-raising. Following in the
footsteps of Jerry Lewis's American model (Lewis and his organization were initially supportive of French efforts, but later on
there was a dispute between them), the AFM initiated the Téléthon in France in 1987. Contrary to many French predictions, the
Téléthon was a great success, raising 185 million francs (about
$35 million). It averaged close to 250 million francs ($50 million)
a year in the subsequent five years. With these significant funds
raised from the public but not the state, the AFM gained an unprecedented autonomy and decision-making power to orient the
genetics research agenda in France.

In 1988, a team from the AFM including Barataud paid a site
visit to Dausset's lab at the Collège de France. The AFM was
already supporting a small research project at the CEPH. Their
guide, in this fateful meeting of personalities and agendas, was
Daniel Cohen. Barataud, who had anticipated visiting yet another
laboratory just like the others, was struck first by the industrial
setting. "We discovered a concentration of machines of all sorts
amid a forest of wires. On the color screen of a computer, I en-

countered for the first time the four letters that would preoccupy me for several years, ACTG. DNA."[25] It is impossible that this was Barataud's first encounter with the four bases (adenine, cytosine, thymine, and guanine) of DNA. Rather, he was encountering a new site of genetics research: with new representational force and actors not cast in the familiar mold.

Daniel Cohen was not the typical scientist Barataud had come to expect. Cohen forcefully argued to Barataud with an urgent enthusiasm that his central concern in the immediate future could not be the myopathies alone. Until there was progress in understanding the human genome, the AFM's investment in research would not pay off. Barataud was moved; he appreciated the blunt language, the ambition, the verve, and the rationale. In their books, both men agree that the other is strong-willed, dynamic, tempestuous, visionary, and difficult. "Between the exigencies of the patients and the impatience of the researcher there was a convergence on a giant project, far beyond the usual norms. A tumultuous friendship was being born."[26] Their collaboration would not be business as usual. Intercession and intervention and sporadic interference became the order of the day.

The first collaborative project between the CEPH and the AFM never really took off. There was to be a systematic screening for markers. This task required the collection of blood from an extensive network of myopathy families. Public health authorities were neither equipped to carry out nor willing to implement such a project. With the project stalled, Cohen made a counterproposal to Barataud that they create instead a bank of yeast artificial chromosomes (YACs).[27] These chromosomes utilized a powerful new vector capable of incorporating and reproducing a much larger quantity of DNA than previous vectors had been able to handle. Cohen quickly saw the possibility of producing a complete physical map of the human genome. Would the AFM finance it? After consideration and consultation within the organization, the answer was "Yes!" "Since we could not localize the genes for the forty dystrophies, we decided that we needed to map the human genome."[28] It is neither exaggerated nor melodramatic to say that this decision, this collaboration, constituted a major novelty, a courageous and visionary act.[29]

A New Model

The AFM created a new model surpassing the older alternatives of state-funded research (with its established hierarchies, priorities, and politics) versus the private sector (largely pharmaceutical companies with little incentive to work on diseases with small constituencies). The AFM invented and occupied a space between the state and the market by dividing "the public" into subsectors. These sectors were then pitted against each other. The AFM was successful in representing the state health care system and research apparatus as callous and paralyzed by its fear of losing its power to special-interest groups. "French biology has a feud with money. In a country without a plan or resources, certain people, and not the least distinguished ones, would have preferred to do nothing rather than something strange that escaped from their power."[30] Barataud emphasizes that the AFM received no financial or technical aid from the government before 1992 even though such aid was publicly announced.

The AFM's model combined cutting-edge research, clinical practice, and patients group activism. Michel Callon and Vololona Raberhariso argue that it was through focusing on the "gene" that the AFM succeeded in linking the specific interests of the dystrophy groups to the larger public interest. By taking the risky decision to support efforts to map the human genome as the best path to discover the genetic basis of the dystrophies, the AFM also contributed to a broader understanding of the genetic underpinnings of health and disease and thereby built a network of allies. Such a strategy enabled the AFM to legitimately claim that they were supporting their own specific interests while arguing that their efforts would produce a comprehensive payoff for society. The "gene" saved the AFM from having to rely in the long term on the solution of "good intentions" [bons sentiments], a fragile base. The AFM can say in good faith that their goal is to ensure a situation in which everyone is equal before disease. Such a slogan resonates profoundly in France.

Their tactics have been pragmatic ones. The AFM has learned through years of experience that so-called rights to health are not easily conceded but must be fought for and won from the state bureaucracies. One tactic has been to confront the state authorities with their "responsibilities"—as with the Téléthon or their alliance with the CEPH—but then to adopt a pragmatic attitude toward future concessions. The AFM has attempted to avoid abstract positions that would limit their ability to act for the good of their members. Thus, they see "industry" as an inevitable partner in the long run because it alone is capable of ultimately providing the therapies toward which the whole effort is devoted. The AFM is highly cautious about industrial alliances and is adamant about its own nonprofit status as an organization. However, their approach has not been to separate "industry" into its own zone but to develop contractual relationships that acknowledge industrial norms without compromising the AFM's integrity in the eyes of the French public as well as their own. Similarly, the AFM has taken an extremely cautious stance toward "ethics."

Généthon: A Factory to Find Genes

At the end of 1988 (or early in 1989), in further discussions with Cohen, Barataud proposed the idea of "a large factory to find genes involved in hereditary conditions, one that would serve all of the scientific community."[31] How, he asked Cohen, would Cohen organize such an installation? Cohen recounts how he returned two weeks later with his plan. The guiding idea would be to create an enormous "factory" to produce genetic information. It would transform the way that such information was produced and the way it was distributed: findings would be made public and disseminated as widely and rapidly as possible. Such an installation would take the core principles of Dausset's CEPH and apply them at a grander scale. In this way, Cohen argued, the AFM would contribute to the general project of genetic mapping, which had the potential to benefit everyone. At least initially, the Téléthon would serve as the core source of funding, thereby enabling the AFM-CEPH venture to move rapidly to an unprecedented scale. "We were hot. The program I outlined marked a turning point in the history of humanity. It would permit us to go ten times faster and at much less expense than what the best American teams were proposing. My megalomania resonated with his. Barataud could only react positively. We were off."[32] Généthon was soon to be a reality.

In his 1993 book *Voyage autour du genome: Le tour du monde en 80 labos,* Bertrand Jordan, a French immunologist, observes that France had made almost no contributions to the series of remarkable technological advances such as the polymerase chain reaction or gel electrophoresis that marked the advent of the biotechnology industry as well as genomic mapping. This state of things is no doubt related, he writes, "to the feeble esteem which technological objects have in France."[33] The French infrastructure was not organized to achieve such results, and during the 1980s and through the mid-1990s, there basically was no biotech start-up industry. Although the research establishment was unquestionably world class in many domains, instrumentation and technology development, as well as rapid reorganization of priorities, were not among its strengths. One reason for this situation was simply structural. In the early 1990s, the budget of the Life Sciences section of the Centre National de Recherche Scientifique (CNRS) was about 2 billion francs ($400 million). Fixed salary expenses accounted for about 75% of this total. A substantial portion of the remaining funds was allotted to fixed-cost items such as equipment, maintenance, and building costs. Jordan estimates that there was at most a margin of perhaps several tens of millions of francs potentially available for innovation. The state of the government medical research bureaucracy, Institut National de la Science et Recherche Médicale (INSERM), was similar. Further, those occupying the places of choice within the research establishment protected their own positions and were certainly not eager to open new avenues of innovation they did not control. In this light, the approximately 100 million francs ($20 million) provided by the AFM for research in genetic maladies represented a highly significant sum.

The nature of the research enterprise in molecular biology and genetics assembled by the AFM and the CEPH differed profoundly in its understanding of who provided the agenda, to whom the scientists were responsible, and how such operations should be organized. The CNRS and INSERM labs are governed by strict rules concerning personnel and purchases and, even in the best of times, have very little margin for maneuvering, even if they so desired. The AFM consciously articulated its strategy in light of what it saw as these brakes on innovation, discovery, and delivery. In an important sense, their model was both much more entrepreneurial and much more civic. Critics rightly point out that the AFM employs significant numbers of INSERM and CNRS scientists—Jean Weissenbach, the scientific director of Généthon until 1997, most prominent among them—and is not obliged to take the overall picture of French research into account in its planning. The CEPH offers an even greater contrast than Généthon. It is a complex hybrid with money coming from Dausset's funds and from the AFM allocations but also from the European Community and granting agencies abroad. Jordan observes that the CEPH is more like "an industrial setting" [une entreprise] than a public laboratory, especially when it comes to personnel matters. In this regard it is perhaps closer to the original genome centers in the United States, whose mandate was competitive technology development. However, less external evaluation was imposed on the CEPH, and its planning was even more open-ended and improvisational than would be possible in an American genome center with its benchmarks and renewable-grant applications to take into account.

Généthon was up and running by the end of 1990. This achievement was itself impressive. In 1992, 130 people were working there, financed in great part (70%) by 74 million francs from the AFM, with the French government providing the majority of the other funds. Barataud and Cohen and their colleagues understood extremely well that communication and publicity would be telling elements in the success or failure of their audacious enterprise. The showpiece of Généthon as an "industrial" site was the so-called Mark II room, named after the rows of equipment that were prominently displayed on an entire floor of Généthon's modern building in Evry, twenty miles south of Paris. The room provided a powerful example of machines, automation, and the industrial division of labor. It became an obligatory stop for visitors and for press photos. As half of Généthon's projects were performed by external teams who used the facilities for specific periods of time to localize specific genes, a large number of outsiders were exposed to this display of machine might. In the CEPH's promotional brochures, Cohen identified himself as the Henry Ford of biology. Whatever the exact importance of the Mark II room in achieving Généthon's goals, it was unquestionably a public relations success.

The accomplishments of AFM-CEPH-Généthon during 1992

dramatically placed France at the center of the world genomic scene. Jean Weissenbach and his team rapidly produced an impressive second-generation genetic map. Cohen and his chief genomics scientist, Ilya Chumakov, whom Cohen had recruited from Russia, startled the Americans by publishing a complete physical map of chromosome 21 in *Nature*.[34] In his own words, Cohen "wowed the crowd of researchers attending the annual genome meeting at Cold Spring Harbor in May 1992, unveiling results on a physical map of chromosome 21 far more advanced than most groups had expected."[35] They threw down the gauntlet by promising a general physical map of the whole human genome. They kept their promise; in a stunning coup, they beat the Americans by the end of 1993. Cohen hired an American public relations firm, who made sure that no one who might be concerned would remain unaware of his and the CEPH's achievement.

3

Field Notes:
The CEPH after Its Victory

Daniel Cohen arrives in his office at the CEPH with a flurry, drops his carry-on bag on a chair, and gives me a warm greeting. He says they almost didn't take off in Boston because of a big snowstorm. He relaxes into a leather chair in front of his big modern desk (most of which is occupied by a large Mac, with a scattering of ficus leaves from the tree in the corner), leans toward me, and we plunge into discussion. He lays out what he calls his "Manhattan project" for the CEPH. Cohen had described his principles and projects in his book *Les gènes de l'espoir: A la découverte du genome humain* (The genes of hope: On the discovery of the human genome; published in fall 1993 and widely promoted). The book is a manifesto on the present and future state of relations among genetics, society, and politics, as well as a summary history of the CEPH and the genome project. No one can accuse Cohen of hedging his bets or being mealymouthed; the book takes strong and unequivocal positions on a wide range of issues. As I begin to get to know Cohen better, it becomes clear to me that these truly were his positions, or perhaps more accurately, these were his positions when the book was produced. One could say Cohen has an experimental relation to things, both emphatic and provisional, a sort of "this is what I think, let's try it out, this is how I see things today, show me that I am wrong," attitude.

Millennium

The reason he travels to Boston so often is that he is engaged in getting a "start-up" company there off the ground. The other main player is Eric Lander, a rising star in the genome world, employed at both MIT and the Whitehead Institute. The latter is financed by private money and has a controversial agreement with a number of leading MIT and Harvard scientists that redefines university/business relations. Financed by a rich developer, Jack Whitehead, the institute has a new building going up in Cambridge adjacent to MIT. In his book, Cohen says Millennium is devoted to finding disease genes with a therapeutic potential. Its strategy is to bring together Nobel Prize winners, famous researchers, and a first-rate staff. The planning meetings at the Whitehead Institute are filled, Cohen exults, with "astonishing reflections." He is sure that tomorrow's genetic therapeutics will be born out of this type of brainstorming. This mode of work fits Cohen's approach to things: take on the most important projects, get the top scientists to brainstorm, and be very, very ambitious about scale. For Cohen, this is the moment of industrialization in the life sciences.

Knowing that such alliances are controversial in France, Cohen was careful to point out in his book: "For a French researcher this is an opportunity. Government officials have encouraged me to take part."[1] Cohen knows very well that getting involved with the privatization of research, especially in alliance with the Americans, is treacherous terrain. He is proceeding in a highly visible fashion, writing about Millennium in his book, consulting with the government, involving key players from the CEPH, talking to the anthropologist.

Tunisia

Cohen's starting point is that "[o]bviously one should not expect any vast humanitarian projects from private enterprise."[2] Even projects with such overwhelming public health implications as a malaria vaccine are seen by the investment community and the large multinational pharmaceutical companies as having almost no commercial potential. For example, work on obesity, searching for a "fat gene," will advance much more rapidly than will work on any number of conditions that are more important by any standard other than a commercial one. Consequently, researchers from rich countries are faced with a real moral problem, a problem they are not addressing adequately.

Cohen is adamant that the CEPH should take the lead on this issue. Specifically, he is already working on constructing a huge genome sequencing facility in Tunisia. He intends to construct, on land already donated by the Tunisian government for the purpose, a world-class facility dedicated to working on the genetics of the problems that plague the Third World but have low commercial value. He is insistent, returning to the theme several times, that establishing such a center is not an act of charity. The financing will be done by the private sector. Cohen has already created a board to give the fund-raising a high international profile; the board includes such scientific luminaries as James Watson. The project is more than paternalism; Cohen conceives of it as an innovative and productive political and scientific action. One of its main goals is to repatriate Tunisian researchers and thereby counter one of the major factors in underdevelopment, the brain drain. Tunisia is, Cohen asserts, the perfect place to begin. Its economic growth rate is 6%. Its demography is under control. Islamic fundamentalism is not a problem. It has a highly educated population. In sum, Tunisia (at the crossroads of Africa, the Middle East, and the Mediterranean) is ideally suited to host such a center. Not coincidentally, Cohen was born in Tunisia and remains emotionally and politically attached to it.

Aging

Cohen takes up the argument in the book that there is growing evidence for a genetic component in the life span of each species. For humans, the life span seems to be 120 years. The identification of several monogenetic conditions like Werner's syndrome, where very dramatic premature aging takes place, indicates that aging processes are strongly genetically controlled. There is also growing evidence that the ends of chromosomes, the telomeres, play an important function in regulating the life span. It has been shown that these regions (whose exact functions in the chromosome are not well understood) gradually reduce in size over time. Given the grave impact on the aging process of certain monogenetic conditions, it is plausible, Cohen believes, that there could be only a small number of genes involved in setting limits on the human life span. In that case, it shouldn't take more than a few years to identify them. Probably two different types of genes are at work: those that provide added resistance to the common problems that attack people in the industrial world between seventy and ninety-five years of age (cancer, heart disease, stroke, etc.) and other genes that might well regulate a biological clock, perhaps one present in all living beings.

Cohen concludes that an audacious research strategy to isolate these genes should be undertaken. The ideal control group exists, the close to 5,000 hundred-year-olds in France (they receive certificates from their mayors on their hundredth birthday). It should be simple to compare the genomes of such people (longevity frequently runs in families) with those with a more statistically standard life span.[3] In 1991, the CEPH had started a project on aging named Chronos. Noting that the hundred-year-olds don't die "gâteux" [dotty] unless they get Alzheimer's disease, Cohen concludes that the norm of life is to retain one's mental faculties into old age. In his book, Cohen notes that if large numbers of people lived substantially longer and healthier lives, there would likely be important social consequences (social integration, employment, and expense). These consequences, he writes, are not his speciality but ought to be given serious attention now.

Social and Ethical Concerns

Cohen says he has been criticized in France for misplaced priorities. Some people wonder why he is working on the genes for aging when there are so many other pressing health needs in society. After some reflection on these and other criticisms, Cohen has decided that these are good questions. "After all," he says, "we are scientists. How are we supposed to know what society wants from us? That is something for others to debate and formulate; we are perfectly willing to listen and to respond." Dausset has always been attentive to ethical concerns. Cohen thinks it would be a good experiment to have a socially or philosophically informed person resident at the CEPH (or a series of such people who visit the CEPH) to advise him and open up public debates both within the CEPH and without. He wants to set up a window of *transparence* at the CEPH. *Transparence* (literally "transparency") has become a catchword, a key symbol, but what does it mean?

I ask him what he expects from me. "Your reflections," he answers as he rushes out. "We will talk."

Cohen has engaged an American public relations firm for the CEPH. That's why they have received so much press coverage for their recent triumphs.

On his desk sits a diplomatic pouch with six copies of his book soon to be on their way to the Vatican. His secretary has written out a variety of dedications, which await Cohen's signature.

The CEPH is having money problems with the French ministries. He says that government officials at the state medical research center (INSERM) do not like him at all because he doesn't play by their rules. Others, sympathetic to the CEPH (or antipathetic to his enemies), point out that France is well known for contempt, jealousy, and settling of petty accounts between the representatives of diverse institutions, both public and private. I point out that such practices and sentiments are hardly absent in the United States.

Cohen works in a disorganized and spontaneous manner. He travels an astounding amount (just back from Boston, Argentina, Tunisia). He does not hold regular meetings. Cohen absolutely thrives on the constant and literal *fuite en avant*.

Eugenics

The gossip at a Parisian dinner party of people in the social studies of science world was that Cohen's book was scandalous (shaking of heads and clicking of tongues) because it brushed close to eugenics. When I asked one acquaintance what he thought about Cohen's specific positions—"Are you really against predictive medicine [the French term for genetic risk evaluation]?"—he backpedaled to a more cautious raising of eyebrows. There is nothing evasive about Cohen's stance. He extols recent advances in genetics as a decisive step toward "the amelioration of our genetic patrimony. Let's not be afraid of words. This will be a form of eugenics but one that seeks to conserve not eliminate. It will be a humanist and nontotalitarian one; that doesn't frighten me more than the discovery of vaccinations, antibiotics, or improvements in childbirth." Not content to let things rest with these rhetorically provocative but basically innocuous comments (very few people attack public health measures), Cohen goes further. He is talking about the creation, not of some fantasy superman [sur-homme], but, rather, of a possible acceleration and guidance of evolution, a superevolution [sur-évolution]. "I believe it will be possible to engender a human that is more complex, more refined, more subtle, farther from animals, than the ones we have today." Citing the French philosopher Dominique Lecourt (a former pupil of Georges Canguilhem), Cohen advocates intervention in the "intimacy of genetic material" because it could increase our freedom.[4]

Lecourt had recently published a book, *Contre la peur* (Against Fear), that was a vigorous defense of science as essential to the fate of humanity. Lecourt identified critics of science as critics of Reason, Enlightenment, and ultimately of the French Republic.[5] Lecourt has written a book about the growth of creationism in the United States to show how dangerous anti-Enlightenment positions have become. Cohen portrays a picture of a deeply embattled scientific community: "For some fifteen years, the little international community of genetics has had to face a kind of intellectual intimidation calling for scientific abstinence, prudence, self-censure, moratoria, in a word a voluntary cessation of activity."[6] He approvingly cites Pierre-Andre Taguieff, a leading theoretician of racism: "It is true that the prophetic French antiscience rhetoric can be best seen as a small and late echo of the great hunt for sorcerers' apprentices unleashed at the end of the 1960s in the U.S."[7] Cohen will wage the "just fight" in France against the fear of science. That fear has "engendered a series of 'machines,' all those ethics committees whose moral competence bizarrely stops at the threshold of biology and medicine; who never are consulted over the fact that the majority of research moneys go to military and nuclear projects; who practice the logic of suspicion, of the slippery slope, and who offer instead an absolute conservatism, a totalitarian control of science."[8] For Cohen, humanity has more to fear from military electronics than from gene therapy. Finally, he wonders: "Why make the natural order as it now stands sacred?"[9] After all, nature alters our genome with each new being that comes into existence; without errors and mutations there would be no evolution.

The one arena Cohen does express caution and scientific doubt about is behavioral genetics. We will not be able to optimize behavior because there is no given norm for behavior. The discovery of the extraordinary variability of our genetic heritage has definitely ruined not only the concept of race but also that of a biological "norm": "The only norm is that there is no norm."[10]

FROM THE INSIDE: DAYS IN THE FIELD, JANUARY 1994

Christian Rebollo, the chief administrator at the CEPH, who will become my invaluable guide and eventually my friend, provides me with an initial overview of the organization from the inside. Rebollo has a *license* (the equivalent of an M.S.) in biology and had worked for ten years in an industrial biology company. Early in 1992, he was brought to the CEPH as part of a reorganization effort to bring the institution out of a crisis that had entered into an acute stage. The CEPH had been attacked in the press for the way it handled its business affairs. It was true, Rebollo said, that the CEPH was underadministered and casually run, but it was not, and had not been, corrupt.

Rebollo is a composed, patient, flexible, and orderly man. He is singularly lacking in pretense. He is observant, analytic, slow to judge, and a very good storyteller. Before too long, I dub him the Albert Camus figure at the CEPH. Rebollo becomes for me a contemporary "l'homme moyen sensuel," "the average man" that Camus extolled.

Rebollo says the phrase Cohen uses in his book to characterize his relationship with Dausset, "on va bien s'amuser" [we are going to have a good time together], is a good description; they are very different but complementary character types. The alliance is a real one, a powerful and fortuitous convergence of interests and styles. He comments that it is even more impressive in that it has permitted a flowering of each man's personality. Rebollo stressed, rather bluntly, but in a matter-of-fact tone, that Dausset is nearing the end of the scientific road. As the French say, Jean Dausset is "un grand monsieur," in the autumn of his life; he is content to function in an institution he created. His role at the CEPH and in the world of science has become more ceremonial, although not entirely so by any means. Dausset is aware that he is aging and frequently will cut short his attendance at events or encounters to economize his energy. How and when he will finally relinquish things, and under what conditions, remain to be seen. Cohen has never crossed him and is always publicly respectful but there is no doubt that Cohen runs the CEPH. Dausset keeps abreast of personnel issues but, at first sight, does not seem to have a major input into decisions over scientific direction and strategies. However, Rebollo insists, the reality is more complex; before making any major decision, Cohen consults at length with Dausset to discuss various options, and the final decisions are the product of those consultations.

Throughout the CEPH's short history, administration had never been the major preoccupation of the management. The resources devoted to administering the lab were always below the real needs, and the tools provided to administer the institution were frankly inadequate. Nonetheless, the administrators, secretaries, maintenance personnel, etc., were highly motivated and had contributed to the dynamism of the CEPH. Fundamentally, the CEPH's resources have been devoted to doing the science. The CEPH has grown in an incremental fashion without systematic planning or typical bureaucratic normalization. This casual attitude toward administration created a very free scientific and social environment. Ultimately, however, it proved dangerous. The CEPH has many enemies and has needlessly made itself vulnerable to attack. The tiny administration makes the CEPH very atypical not only for French science but for French industry. The administration was weak and not respected by the CEPH scientific personnel. Cohen and Dausset now understand that its authority needs to be reinforced both internally and externally. The challenge is to meet the state's norms but not fall into the stifling situation found in so many other labs. The AFM, extremely vigilant about transparency, is doing an admirable job on this level.

The CEPH, in part due to the changes in its juridical status (it became a "foundation" on March 15, 1993), is in the process of reinforcing its management, normalizing its internal procedures, and attempting to improve its management criteria in general.

Cohen chooses people not according to their titles but according to what he sees as their potential and their immediate utility. When the AFM money started coming in, Cohen essentially put out advertisements and simply hired people as they applied. There has been surprisingly little working of traditional networks or repaying of favors through hiring. The CEPH's philosophy has been the "management of skills, not degrees" [gestion de compétences et non de titres]. Actually, Rebollo says there is a strong tradition in French industry of what he calls autodidacts. He cites studies indicating that in certain sectors their presence is very high; they are crucial during the early stages of an organization's growth for the innovation and the motivation they provide, but when the inevitable "plateau" arrives, another type of administration is required. Rebollo himself, of course, is as much an autodidact in administration as Cohen is in genetics.

Cohen and Dausset are committed to preventing the famous French handicap of resisting innovation from taking place at the CEPH. Cohen draws heavily from his own experience, which explains his nearly incessant activity, his frequent trips, and his extensive network of personal contacts. Cohen trusts his own experiences and intuitions; he is synthetic, conceptual, and audacious. It is fair to say, "Il ne s'entoure pas de forme" [the man operates as he sees fit].

The current moment is one of responsibility of the lab heads [les chefs de labos]. Cohen has applied what he calls a "vertical management style," which is to say that although there is a certain overriding CEPH spirit, there is no organizational structure that brings the whole organization together very frequently. Cohen is the obligatory contact point for each of the lab heads. This structure has led to complaints because it is frequently very hard to get to see him. This irregularity imposes extra burdens on the lab heads. Although all major decisions must pass through Cohen, he generally does not micromanage. However, he is perfectly capable, when he sees fit, of spending entire days with those responsible for specific programs that are currently important to him while neglecting, even for months at a time, other projects that he judges to be of secondary importance, or at least not current priorities. Hence, the operation is a personalistic one: open, dynamic, sponta-neous, tempestuous. Rebollo's challenge is to maintain this open style of operation while putting it on a solid footing. Dausset and Cohen are at the top. Then come the lab heads, who are responsi-ble for defining their scientific projects, managing their teams, and respecting their budgets. The CEPH has no union structure although there are a number of active in-house committees for such things as safety matters.

Robots and Automation

Patrick Cohen (no relation to Daniel) is in charge of the robots. He is an extremely accommodating, soft-spoken man, in his early thirties, with a background in biology. While working on a dys-trophy project in a hospital in the Paris region, he saw an an-nouncement of an opening at the CEPH. Because his hospital job was becoming routine, he went for an interview. Although he had not felt like he had performed well at the interview, a week later he was offered a position and took it. Patrick Cohen is a typical example of Daniel Cohen's hiring of people with limited establish-ment credentials and offering them positions of responsibility. Patrick Cohen has always had a "passion" for machines, but he had no real experience in automation in molecular biology. He is loyal to and dependent on Daniel.

In principle, Patrick Cohen is available for consultation by any of the teams at the CEPH. In reality, he works with only two of them (Howard Cann's team, who work with the "CEPH families," and the YAC library). The work keeps him quite busy. His role is a kind of translator-facilitator: he does not build the machines although he does make modifications based on his own biological experience or that of the teams for whom he provides service. An engineer from Généthon is responsible for the basic technical maintenance.

Patrick shows me two "robots," and although they are func-tional, this is not the stuff of futuristic science fiction. They are laborsaving devices; they perform tasks such as filling the wells on plates and then transferring fluid from those wells to other wells. The machine does the work of one or possibly two technicians as well as freeing the scientist to do other things. Initially, there was quite a bit of resistance to the machines, especially because it meant coordinating scientific work with machine time. People are now used to such things.

Patrick Cohen shows me how he prepares the DNA of the "CEPH families." He uses a process that keeps the DNA very stable. The samples can now be sent through the mail, resulting in cost reduction and ease of use. Machines and their routines are now simply part of what a genetics facility is. The actual technical innovations at the CEPH are not startling. The innovation really consists more in the way the machines have been integrated into the practices, how they have changed authority patterns, and how central they have become to achieving goals.

Patrick Cohen, like everyone else I talk to, is clear that the present core projects of the CEPH (the distribution of YACs and the mapping) will not carry them indefinitely into the future. Consequently, there is a floating incertitude about what form future research and organization will take. Everyone awaits Daniel Cohen's decisions. Patrick, at first timidly but then more emphatically, says that Millennium will play a central role. Although the CEPH is well financed, there is constant insecurity about future funding. There is a good deal of jealousy, envy, and dislike of the CEPH in the state agencies. They cannot be counted on for much support. In France the divide between the industrial milieu and the research world is important; French industrialists in the pharmaceutical world have not come to the CEPH with projects. Everyone knows that new sources of money will have to be found in the United States. By establishing contacts with Millennium, Daniel is saving the ship.

Gene therapy might prove to be another future direction but it is too early to tell. This remark leads to a discussion of bioethics. Patrick is very much in favor of the current plans for legislation. Is he troubled by projects at the CEPH? "No." However, he knows of work on gene transfer going on in bioagricultural labs and such techniques can be appled to humans. It is not being done but it could be. In sum, society is not to be trusted; legislation to set limits is necessary.

Le Projet Chronos: Eu-Genetics

The "aging" [vieillissement] project is looking for genes that protect people from the "ordinary," in the sense of statistically expectable, maladies. The project is distinctive in that its goal is to identify the conditions for normal health and longevity; this work is not searching for a mutation or a disease gene. The question is: why do some achieve such longevity and others don't? Perhaps there are genes that offer "extra" protection against assault. (1) What are the species factors involved in aging? For example, why does mouse tissue age fifty times more quickly than human tissue? (2) What is the genetic basis for different life spans in humans as individuals? The first question can be answered at the cellular level by exploring the genetic basis of metabolism. As the mitochondria (spherical or elongated organelles in the cytoplasm of nearly all eukaryotic cells that contain genetic material and many enzymes important for cell metabolism, including those responsible for the conversion of food to usable energy) are the source and transfer point for all energy transactions in the cell, it is logical to look for the genetic basis of that activity. A great deal of progress has been made in recent decades on establishing the number of mitochondria per cell, the activity of mitochondrial enzymes, the physiology of cell respiration, its operation under stress conditions, etc. The second question is appropriately taken up by the inverse genetics approach used by the CEPH. As François Schachter wrote in the CEPH's internal newsletter: "The method of inverse genetics has already proved itself useful in the study of multifactorial conditions such as diabetes and cystic fibrosis; this is the first time that the method will be applied to the analysis of a nonpathological condition: longevity is antipathology."[11] In the course of mapping the human genome, the CEPH and its collaborators have developed many polymorphic probes and there are a variety of candidates to test.

The CEPH project is divided into two studies. The first deals with the hundred-year-olds, "les Centenaires." In addition to consulting state records, the CEPH team also places ads in newspapers and other media. Its toll-free telephone number has generated a good response. Because there are so few people in this cohort, it is hard to design a definitive study. The other study concerns brother-sister combinations in which the brother is at least ninety years old and the sister at least ninety-five. These are more complex linkage studies of a classic sort. However, what genes to look for is far from clear, and the description of desirable gene markers is very preliminary.

In a strict sense this project could be characterized as "eugenic" in the sense of "good genes." In fact, historically, eugenics projects in France were directed at improving public health, not racial improvement or elimination of "inferiors." The crowning glory of French eugenics was mandatory prenatal examinations.[12] And of course previously there was no science of genetics that was capable of identifying genes. Perhaps it is better to say that the CEPH project is a biopolitical one, operating in the name of the health and well-being of the individual and the collectivity.[13]

Project Chronos is partially supported by the Ibsen Foundation. The Ibsen Foundation is funded by the director of a pharmaceutical company who is known to support far-right groups. Such associations have reinforced suspicions about the CEPH as being a tool of the forces of eugenics or capitalism or both.

François Schachter's parents came to France from Romania. He has "always" been passionately interested in aging. In the past, he was actively discouraged from pursuing the topic; when visiting a large number of labs, he was repeatedly told "perhaps in five years but not now." In the United States he attended a number of conferences where there was growing interest in apoptosis, programmed cell death. When he met Cohen in Paris, Cohen strongly encouraged him to pursue this line of work. A second meeting some time later produced an offer for Schachter to come to the CEPH. Despite Schachter's lack of formal training or credentials in the field of research, Cohen provided minimal resources and proposed that they just see what he could find. The CEPH resources are still relatively limited, but with the pharmaceutical funding they have an active lab. Schachter, who is somewhat shy, does not have a lot of contact with others at the CEPH. He knows very little about the other operations at the CEPH and not much about the genomics world either. He is basically a cell biologist passing through an open genomics door. He fits Cohen's mold of someone who is not overly specialized, is relatively marginal, and is motivated by a vision. That combination and machines, in Cohen's view, is how breakthroughs are made today.

Schachter hopes that in thirty years it will be possible to envisage the mastery of aging. He doesn't explain what that might mean. He thinks most of the ethical debates in the genetics arena are interesting but not really connected to what is feasible today in terms of current technology. He feels that a scientist should be attentive to ethical discussions. However, so far, he has not encountered any ethical problems that would impede his research. I asked him if he thought patenting is an ethical problem. Schachter responded that the issue had not come up much so far at the CEPH. He approached Cohen at one point and they discussed it but they did not pursue the issue after that. Schachter can see no a priori reason for not patenting their work, especially if and when they find important genes involved in the aging process, but so far they hadn't made major discoveries or patented anything.

When asked how many genes were likely to be involved in aging, Schachter replied, "Many, perhaps one hundred." These genes are probably not directly causal but operate in complex ways with the environment that we don't understand yet.

Several days ago, Cohen asserted that Schachter was unemployed when he came to the CEPH. Schachter is beginning to think of the work on aging as his project; it is going to his head. There is more than a hint of a threat in the way Cohen announces this in public.

On January 26, 1994, a crew from North German Television comes to the CEPH to film a portion of a program they are making on aging. The interviewer is a late-thirtyish German dressed in wide dinosaur tie, yellow wool shirt, cowboy belt, and yellow-rimmed glasses. He is very nervous and very polite. He has a cellular phone. He says he is of Russian-Polish-Italian stock. He also says he can trace back his family eight hundred years. The ironies of a German TV producer (North German Broadcasting Company) interviewing two Auschwitz survivors on genetics are not lost on him. The subjects are the Klein brothers, animated and alert at ninety-four and ninety-seven years old. They tell the interviewer that they had had a miserable, impoverished childhood. They have survived Auschwitz, as well as the flu epidemic of 1917.

The first question is whether or not they have a recipe for

longevity. "Well, sort of." They have always lived prudently [justement], no excess. They have never smoked. They take long walks. They eat moderately. The most important thing is having the will to live. What counts is having the will to adapt to whatever circumstances come along. One must not accept death. The next question: "Do you think about genetics?" The answer is "no."

Cell-Imaging Technology

Dora Sherif's research interests are in cancer but her time is devoted to improving cell visualization technology. Trained in Tunisia and France in genetics, she has been earning her living doing work on in situ hybridization, marking her more as a technician than a scientist. She ardently wants a regular position at the CEPH, which has been paying her salary for the last three years although she is still officially attached to the INSERM lab. She needs to be part of a research project; otherwise she fears being typecast as a technician. Cohen's promises and absences add to the uncertainty of her situation.

We have a long discussion about Tunisia. Sherif is from the upper bourgeoisie and well connected. People cast Cohen as a stereotypical Tunisian Jew, a group for whom appearances, *le paraître,* are of great significance. She has great nostalgia for Tunisia and a strong sense that the Tunisian Jews, like Cohen, are as much Tunisian as anything else.

Sherif agrees with Cohen that Tunisia is a modernizing country. The president, Ben Ali, hesitated over how to deal with the fundamentalist opposition in his first year in office but then intervened in a strong and decisive fashion, crippling them as much as possible. These attacks were effective, but they also implied that a larger, secularist, modernizing policy had to be put into place to ensure that things remained that way. Hence, technocrats and liberals are now in place in the key ministries. Modernization is a core symbol for Tunisia and Tunisians.

Cohen was received as a head of state by the Tunisian government, with red carpets, motorcycles, piles of presents, and a major diplomatic reception as well as extensive television coverage. This show was a governmental commitment to the prodigal son. Cohen was deeply moved and is committed to doing something important in Tunisia.

A theme echoed by everyone is that Cohen's absences are getting to be a major problem. No one knows where the CEPH is going and what he intends to do. They are unable to talk with him. It is very frustrating. He is just throwing out project ideas to see what will happen. This leads to a lot of discussion about what it all means and why he is doing what he is doing. Low information equals high gossip. Sherif's example is the Tunisia project, in which it is clear she should have a part but has no idea what that will be.

Between Things

Rebollo is adamant that the CEPH is in a major moment of transition. The next three months are absolutely crucial. The future of the institution is at stake. The first sign of this danger is the refusal of the ministries to allocate more money than the 33 million francs originally budgeted. The signal concerns more than the money itself. Cohen is being rebuked. The CEPH is fragile. There is definitely the possibility of a domino effect: if one part collapses, other parts could fall as well. So much tension and resentment have built up both within and without the institution that it could not stand a strong shock.

Cohen, Rebollo thinks, could well leave! It is not his style to stay and be humiliated. Rather, it is perfectly conceivable that he is concocting other plans (hence all the absences). What these things are is unclear. When I say that Nobel Prizes and money are found in the United States, Rebollo agrees; when I say that I think Cohen could only flourish in France, Rebollo disagrees.

The current budget deficit is a serious one. Unless something changes, it will inevitably lead to layoffs (as many as three scientists and four technicians, hence a whole team). The older methods of filling budget gaps by means of creative bookkeeping are no longer tolerated by the government. Further, to compound the situation, there is a conjunctural tension, or minicrisis, with the AFM. The 15 million francs they have contracted to give the CEPH is in doubt or at least under discussion. Should Barataud decide to withdraw this money, it would be catastrophic for the CEPH. Rebollo says that Dausset is very uneasy, worried about the future. Any serious blow to the CEPH could be mortal for him.

Statistics

Mark Lathrop comes across as a mixture of a classic nerd and an Anglo-Saxon "real man"; he sports the black beard and shaggy hair look that many North American scientists adopt. There is widespread respect for him at the CEPH. He is seen as one of the pillars of the CEPH's glory days, having introduced sophisticated statistical methods and advanced computational techniques, as well as successfully advocating industrialization of gene hunting. He is moving to Oxford, where he will set up a new center for gene mapping, partially supported by private money.[14] Lathrop, it is said, could represent substantial competition to the CEPH and Millennium because he is highly efficient, organized, and intelligent and spends little time talking. Cohen expresses little concern.

That afternoon there is a talk by a "hot shot" from the University of Michigan on multifactorial disease modeling. There is a very large turnout. He is Lathrop's friend and is consequently accorded respect. He is dressed in a cheap blazer, blue shirt, and blue tie, with gray pants and functional shoes. His beard is trim. His talk is in English. He puts up transparency after transparency and then reads word for word what is on the transparency. Speaking with one's back to the audience is typical of many scientific talks. At the end, the questions from the audience vary from the skeptical to the hostile. The audience is composed of people currently working on gene identification, and the professor readily agrees that his models are not applicable to present-day situations. They have no immediate utility. He is certain that the future will be different. Lathrop apparently agrees.

Cancer: Les Belges

Rumors circulate that Cohen has formally invited two Belgian cancer specialists to join the CEPH. As Cohen told Le Monde, he has been looking for cancer projects for the CEPH. The general reaction in the hallways and smoke-filled lunchroom, with its microwaves humming, is negative. No one seems to have been consulted. As these Belgians are being given space and money, both in short supply, there is a strong sense of resentment. The comments on their research proposal, which has somehow circulated, are largely negative if vague. Some of these evaluations have been passed on to Dausset. The Belgians are quoted as saying at the elevator that they have come to the CEPH to discover a cure for cancer; then they will go on to other things. Much derision is directed at them.

A lab head apologetically postpones our follow-up talk because she has been abruptly informed that part of her lab space is to be transferred to the cancer project. Consequently, she is coping with moving part of her team and equipment into smaller quarters. She looks demoralized. She has not protested the decision to Cohen.

Cohen takes a key person aside and tells him that the proposal that is circulating is a sham. The Belgians—their names are Adam Telerman and Robert Amson—had written it poorly at his request so as not to reveal too much of their strategy. They have excellent letters of recommendation as well as letters of serious offers from industry. Cohen sees something important in the Belgians' work.[15] He makes his support for them totally unequivocal to everyone. He is giving them a total vote of confidence. Jacqui Beckmann, a leading geneticist at the CEPH, comes from the same Jewish networks in Antwerp that Adam and Robert do, except there is a major class difference between them, as the latter are connected to the rich diamond merchant families. (Later they assert that it is Beckmann who comes from the richer family.) At different times, all three of these Belgians have worked at the Weizmann Institute (Beckmann worked on an agro-biotech project as well). Beckmann came to the CEPH in 1990 to work with Lathrop and ended up staying on. The Jewish bond is strong. Cohen, Froguel, Beckmann, *les Belges,* and others explicitly mention it as a tie among them and as one reason they accept me.

A late afternoon meeting is scheduled with a representative of Rhône-Poulenc, the French-owned multinational pharmaceutical giant that has been the main player in France in the biotech arena (without great success). Telerman gives a lucid presentation for this man, with Cohen, Amson, and myself present. Cohen is clearly interested in the potential of the technology they are using. He is looking for new DNA gene expression technologies which would proceed on a CEPH scale, concentrating not just on single oncogenes or antioncogenes (RAS and P^{53}, the two current stars) but on a very large array of such genes and the multiple pathways that might exist. The idea is that focusing on these genes misses many other things. The Rhône-Poulenc man was quite interested. He was sharp, quiet, listened carefully, and asked questions that synthesized the conversation. He took notes, posed some objections, but offered little overt reaction. The session lasted almost four hours. If Cohen could establish a French funding connection, it would allow him to counter the charges that he was selling out to the Americans. It would also allow him to enter into venture relationships with other companies, American ones.

Telerman and Amson were clear, knowledgeable, and logical. They expressed strong reservations about industry relations. When I asked if there was anything confidential in the session, they said "no." They were aware that their ideas could be stolen, but ultimately for them the point was to cure cancer.

Later in the afternoon I go downstairs (the CEPH has the building's two top floors and two sub-ground-level floors; INSERM cancer researchers occupy the two intermediate floors). One floor down, the same architectural space offers a complete contrast with the CEPH. The space is neat, clean, and quiet; equipment is spartanly placed on the lab benches. Only a few people are present. The offices are large. The offices at the CEPH have been subdivided to increase bench space; leading researchers, for example, work in a cubbyhole that can hardly fit the two people working at a large computer.

The cancer researcher I am introduced to, Fabian Calvo, is cordial and patient. He outlines in a quiet, didactic manner why he believes that miracle cures for cancer are unlikely. He thinks cancer is many things. Most cancer studies (he works on breast cancer) concern later stages of the disease, where one sees the end product of a cascade of other events, preconditions, and cellular and genetic aggressions. The problem is that only recently has the technology been available to allow work on cancer cells drawn from human patients and not laboratory cell lines. He adds that in France there have also been problems getting the doctors' cooperation to provide cancerous tissues, although the situation is now changing. Normal tissue is readily available from plastic surgeons doing cosmetic operations. Only recently has it become possible to do research on cell cultures exhibiting earlier stages of tumorogenesis. Calvo is very interested in techniques that promise to identify multiple genes as they are sequentially expressed (or not expressed) in different tissues.[16] Such information could well provide a greatly enhanced understanding of cancer as a cellular process. His calm and clarity are soothing.

The Lab Heads

Christian Rebollo calls a meeting of lab heads. He informs them that this time the fiscal crisis is real. The ministry wants their budgets strictly in order. Rebollo distributes a memo from Cohen, countersigned by Dausset, requesting an "inventory" from each team. Beyond the fiscal constraints and increased governmental scrutiny, the imminent departure of Mark Lathrop and his team for Oxford raises the question of exactly what material and equipment belong to the CEPH and what belongs to individual researchers. The question really amounts to asking, Rebollo remarks, what exactly is the CEPH's patrimony?

Although Lathrop's departure is the occasion for this discussion, there is no real controversy about him. He has the respect of all at the meeting and no one accuses him of any wrongdoing. Lathrop takes the position that everything on his computers belongs to him because he had raised the money for his projects. No one contests the specifics but there is a good deal of discussion about the general point; a consensus exists that there needs to be a clearer policy about intellectual property. Although everyone present is in favor of having clearer policies and of discussing them more fully in a timely and democratic fashion, there is a common expression of concern that the CEPH operates only through crisis. There are simply too many sudden deadlines and never enough time for sufficient preparation. There is an aggrieved and annoyed unanimity that Dausset and Cohen have been negligent in not paying sufficient attention to planning. They should have set a policy in place long ago: "We need a method instead of always playing the firefighter." Rebollo agrees with all the concerns but reminds everyone that they must deal with the situation at hand. Almost in passing, someone asks: "What is an inventory?" People agree that (1) it cannot simply be a list of what is in the refrigerators, and (2) if anyone wanted to hide anything, it would be basically impossible to stop them. Therefore, an inventory could only yield a sense of the projects under way, the tools developed, and the amount of the work carried out at the CEPH. The real question for many at the meeting is the perpetual crisis atmosphere. The general state is mistrust; hence, things like inventories become crises rather than normal bureaucratic procedures. Toward the end of the meeting, Rebollo says, almost in passing, that the lawyers are drawing up a contract that spells these things out and that everyone at the CEPH will have to sign it. His remark is greeted with silence; everyone had been calling for regular and explicit procedures, but they also realize that once again they have not been consulted on a new policy. The brute fact of this fait accompli takes the steam out of the conversation. Rebollo reiterates his conviction that it is perfectly possible for the CEPH to achieve a clear and transparent structure of accounting and reporting, if everyone cooperates.

Jacqui Beckmann expresses his irritation at being cast as a *casque bleu* (the blue-helmeted United Nations peacekeepers in the Middle East) between warring individuals and groups. The CEPH's good soldiers, team players, are feeling frustrated; they do the work and then others come in, get space, and get Cohen's attention. Cohen needs to act more responsibly. He must stop belittling people. Cohen, Cohen, Cohen.

A lab head laments that there is too much change and too little consultation: "Daniel thinks that any five people he happens to bring together would be able to bring his idea to fruition." The lab head pauses and with a slightly bitter chuckle says, "Perhaps he is right, who knows?"

All these remarks underline one of the most curious dimensions of the CEPH. It is not uniformly staffed with great scientists, towering intellects, or generally even people of great scientific accomplishment. Especially at the lower levels, there is competence but no more than that. The majority of its scientists and technicians have not been recruited from France's elite institutions, and yet, despite it all, despite the fighting and bickering, the lack of structure, so far things have worked out. Cohen's ideas and mode of organization have proved to be extremely fruitful. Everyone agrees, however, that such an organization needs reform.

Sherif is at her wits' end. Although she is technically a CEPH employee, she is currently not attached to any project nor does she have any lab space. She is being paid but her work is completely displaced. She and Calvo had seen Cohen at seven last night (he was two hours late). He simply is not interested in the project that they proposed. The CEPH's cancer slot is for *les Belges*. However, he did not say "no."

The frustration level in the YAC lab is high. Several of the technicians complain that they would find it more satisfying to work on a project from beginning to end. Currently, their work is alienated. There is little chance for them to learn more. They are unhappy at what they see as the overload of work without the satisfaction of experiencing the completion of a whole project. They claim to be, and seem to be, a very productive unit, but others have greater resources and less work to do. They do not feel appreciated.

Later that day there is a discussion about "feminism" in the "aging" lab. There are five women technicians in the lab. I ask if feminism is an issue at the CEPH. Everyone concurs that the "North American style of feminism" would never work in France. One woman says she has Canadian in-laws who tell her that women in Canada refuse to let men open doors for them; she finds this ridiculous. In France, women want to be accepted for their accomplishments. "Is the French system fair?" She responds, "I suppose so," clearly never having thought of it in those terms. "Do you think that Françoise Barré-Sinoussi should have received more credit for her work in the discovery of HIV?" Everyone thinks that she should have, but for them this is less an issue of feminism than of the fact that in France hardworking laboratory people rarely receive appropriate credit and recognition for what they do. There is no resonance at all on the feminism issue. I let it drop.

The South

Mariano Levin, an Argentinean whom Cohen met at a conference in Buenos Aires and invited to work at the CEPH, arrives one morning only to realize abruptly, and to his dismay, that Cohen had not informed anyone at the CEPH about him. Fortunately, several scientists and technicians at the CEPH (and the Institut

Pasteur) find his project on sequencing the genomes of parasites interesting. They form an ad hoc group to help Levin.

The group agrees that the real aim of the study of parasites is to understand their amazing genomic plasticity. So far the recombinant revolution has not impacted the specialists working on parasites to any significant degree. This is partially because the standard genetic techniques that are applicable to bacteria do not carry over to these parasites. They are very plastic organisms which adapt very quickly to changes in the host milieu. The genetic mechanisms for this plasticity remain entirely unknown. The hope is that the genomic mapping approach will shed light on the novel structure of their genomes. It is known that these genomes contain a great amount of redundancy with many repeated sequences and multiple copies of genes. These features seem to be a resource pool available to the organism for rapid adaptation to changing conditions. There is agreement that although the science is exciting, the short-term public health applications are not obvious.

At a lunch with other Latin American scientists, there is a passionate discussion about the politics of the World Health Organization as well as the general state of unequal exchanges between the rich nations of the north and the poorer ones of the south. There is general agreement that not only are the resources vastly unequal but there is insufficient reciprocity, bordering on exploitation. The North American stars almost never establish equal collaboration with their South American colleagues. A Venezuelan scientist, a mild-mannered, contained, extremely lucid and intelligent man, gets quite agitated about the work done in Venezuela on Huntington's chorea. He describes how after a Venezuelan scientist mentioned at a scientific meeting that there was a large population with Huntington's in Venezuela, teams of Yankee scientists descended to take blood and genealogies from everyone in the village. In return, they gave milk and T-shirts to the villagers. For the Venezuelan scientists there was nothing, no significant collaboration. Remarks from the Yankees that the Venezuelans are lazy, etc., have produced great bitterness. The Yankees then go around the world showing pictures of ragged children and the wonderful work they are doing.

Maurice Moncany, the new AIDS specialist, rotund, extremely cheery, bounces into Cohen's office, where I am now centrally installed, and announces, "Je te salue." Following the new political line of supporting the French language, the technocrats in the government, who have just announced a quota on the number of English words that can be used in advertising or elsewhere, are, Moncany proclaims with wit and erudition, simply ignorant that many of these English words came from French in the first place and that their police actions are stupid. He speaks enthusiastically of the rich exchanges between English and French in the twelfth and thirteenth centuries, which diminished only after Jeanne d'Arc. He is immensely chatty, *un bavard,* and hence at times it's hard to get on to other things. He has just come back from Japan, where he spent a year and a half. He went there because the work he was doing in a prestigious French lab was not appreciated. His boss had told him not to publish his results. Moncany offered to wait a year, and if his results stood up to scrutiny, he would leave the lab and publish them, which he has done.[17] Moncany said that scientists are a strange group who spend their lives subjecting themselves to the criticisms of others.

4

Life: Dignity and Value

Biopolitics . . . is the politics of the spiritual animal
kingdom.

FERENC FEHER AND AGNES HELLER, *Biopolitics*

The genome is not sacred. What is sacred are the values linked to our conception of humanity.

ANNE FAGOT-LARGEAULT, "Respect du patrimoine
génétique et respect de la personne"

Since World War II, French bioethical discourse has had two major (overlapping) articulations that form the milieu in which genomics came to be problematized as a social issue. The first was
the system of blood donation. The second emerged in the 1980s
around issues of artificial assisted procreation. The principle of
the "gift" which in France has come to mean "benevolence," that
is, anonymous donation, arose out of the Resistance during World
War II. The *bénévole* complex was a paradigmatic form of how
body materials, social solidarity, public health, and money should
function together. The system maintained an impressive legitimacy for decades. It took the AIDS epidemic and the crisis of
contaminated blood, in which hemophiliacs and others were infected with HIV from blood transfusions, to bring some of the
system's latent tensions into view. The French (or any other large-
scale) system could not operate on benevolently donated supplies
of blood alone, nor was it as exempt from the laws of the interna-

tional market as the citizenry had been led to believe, or from the dynamics, statics, and politics affecting any large-scale bureaucracy. The crisis did not so much undermine the system per se; rather, it highlighted some of the previously hidden assumptions without resolving them. The limits of any "national" system were underscored by the growing internationalism of markets for blood products spurred by advances in therapies for hemophilia arising from new bioengineered products like Factor VIII, an effective clotting factor. Massive blood supplies were required to produce this product, which yielded significant improvements in the quality of life of the hemophiliac population. National symbols, public health practices, and therapeutic efficacy stood in tension, on the verge of disaggregation.

France's National Ethics Committee, formed by a decree of President Mitterand in 1983, was the world's first national ethics body. Although the committee's constitution explicitly denied it legislative powers, it did provide it with a mandate to stimulate national debate and reflection. First, the committee rapidly focused its attention on amorphous concerns over new reproductive technologies (especially in vitro fertilization) and turned those concerns into a paradigmatic ethical crisis. The crisis was made to be paradigmatic in part because there were no guidelines in place from the past to direct the regulation of what the committee succeeded in defining as "new" threats. Thus, articulating the need for regulation was one of the committee's first achievements. Second, the committee linked the debates to the Universal Declaration of Human Rights. The distinctiveness of the French committee's articulation can be seen if one compares it with the situation in the United States, where debates around new reproductive technologies linked them to the philosophical and political struggles over abortion. The French debate did not underscore the link with abortion or with issues of women's right to control their bodies. Rather, the committee rapidly adopted an agenda of articulating universal ethical principles, especially those relevant to defending the "dignity" of the "human person." This meant, as befitted its mandate, not only undertaking (often abstract and inconclusive) discussions of what the human person was but, more concretely, identifying which groups or "interests" were threatening the dignity of the human person. The committee was instrumental in transforming France's official ethical mood from a proud affirmation of acts of benevolent giving to a defensive one requiring vigilance against transgressive threats. The committee adopted a "civilizing" role in the sense in which it was elaborated by Norbert Elias,[1] emphasizing the need to curb, domesticate, and extirpate the vectors of violence that science and capitalism were spreading in the world.

It is within this milieu that debates about the social consequences of mapping the human genome as well as how to configure finances and the production of knowledge about life-forms took place during the late 1980s and early 1990s. While the ethical vector of the national committee and the penumbra discourses spreading around it have been cast more or less explicitly in the "civilizing process" vein, there are other, more muted subthemes and leitmotifs that are not captured by this secular and sovereigntist rhetoric. For a start—not surprisingly, as urgency and crisis are the foundations of the committee's mandate and its mode of self-stylization—the official French ethical discourse does not adequately account for itself. First, the reason so much weight is attached to these issues and pronouncements when they are seemingly self-evident (in sum, the "lessons" of the Nazis) is, I argue, actually far from obvious. French bioethics discourses and institutions, although cast as universals, are only one mode among others of problematizing the issues addressed. However, since the committee speaks in universals, it tends to elide the status of its own particularity. Second, the relations of money and science to solidarity within modernity have been thinly problematized in French social thought largely because it is believed they have been normatively solved. The long-standing and frequent recourse to the state to adjudicate and regulate such matters is of no help in situations where the nation is not the base unit. When it comes to science, capitalism, and ethics, there is much that a discourse of ethical universals and sovereign power fails to regulate and, more pertinently here, lacks the means to conceptualize. For example, discursively ruling that the body is "the person" and both are "outside the market" [hors du commerce] is a stance that is not even consistently applied within French state institutions. It will need a great deal of elaboration if it is to provide a basis for articulating an ethics in a world of genomics, multinational pharmaceutical companies, and militant patients groups.

PARADIGMS SHAKEN: FORCES OF
SCANDAL IN TWO NATIONS

From summer 1982 to spring 1985, blood supplies were being contaminated by the HIV virus (among others) while authorities in all of the Western countries (as well as Japan) hesitated.[2] At issue were different courses of action: those concerning the donors of the blood; those concerning the blood itself; and eventually those concerning what steps to take with existing stocks of blood. Diverse factors were involved in evaluating these courses of action. One factor was the changing scientific understanding of what exactly was causing this new epidemic. Another significant factor was the technological ability to effectively identify potential causative agents or their surrogate markers; subsequently there was the technological question of how to cleanse the blood of contaminating elements. A third set of factors concerned the institutions responsible for taking action—the way they were organized, legitimated, and financed.

The science, the technology, and much of the blood circulated across national borders, across oceans. The institutions and their associated cultures and symbols did not. Specifically, the United States and France had quite different systems for collecting and distributing blood, that most symbolic and sacred of substances in the West. In France, the two systems are often presented as constituting opposed models of public health, collective action, and the state. In France, for almost half a century, these two systems have been represented as icons of two different societies—societies, defenders of the French system underscore, with profoundly different values. These collective representations—including French representations of *les Américains* or, in more racialist terms, *les Anglo-Saxons*—continue to play a significant role in France. In the United States, representations of France, to the extent they exist, were an inconsequential factor in these events. Representational relations are unequal and asymmetric.

The brute fact is that both the French and the American authorities (as well as those of most other countries) failed to take timely and effective action. Although American authorities eventually (May 1985) took significant steps to "protect the blood supply" before the French authorities did (October 1985), clearly some steps could and should have been taken earlier. Again, it is of ethnographic interest that the self-interested and initially reckless conduct of the diverse actors in the United States did not result in an *affaire,* a set of events focused by a narrative. In France, belatedly, the "contaminated blood affair" unfolded slowly at first—beginning in 1987 with calls by hemophiliac groups for justice, followed by several years of judicial maneuvering, leading to the eventual indictment of several highly placed officials of the national transfusion center, and culminating with an explosive 1991 trial given extensive media coverage. Significant questions could be, and have been, raised as to whether the French affair focused the public's attention on the most significant issues. Perhaps one affair covered over another. It is at this level, it seems to me, that an ethnographic perspective becomes pertinent.

Scandalous Facts
During the summer of 1982, clinicians in New York City's Bellevue Hospital publicly made known their suspicions about possible contamination of the blood supply by an unknown agent believed to be causing the new syndrome. They submitted a scientific report to the prestigious *New England Journal of Medicine* and the *Annals of Internal Medicine.* It was rejected. "The editors [of the *New England Journal*] feared a faux pas and the wrath of powerful institutions."[3] The authors were told they lacked sufficient proof and their report would cause alarm. At this point the scientific, medical, and activist communities most involved in the epidemic were sharply divided. Six months later, opinion was shifting. By January 13, 1983, a *New England Journal of Medicine* editorial advised, "physicians involved in the care of hemophiliacs must now be alert to this risk."[4] This advice was not heeded.

Randy Shilts, in his book *And the Band Played On: Politics, People, and the AIDS Epidemic,* describes a key decision-making meeting held at the Atlanta offices of the Centers for Disease Control (CDC) in January 1983. Basically, representatives of all the major interest groups either took no position on screening the blood supply or opposed it. They did so for different reasons. Although there were no tests for HIV at this time, there were commercially available tests for hepatitis B, a virus epidemic in the gay population (rates as high as 80% of those tested) that posed a serious threat to anyone receiving blood. Blood bank officials strongly opposed screening, citing cost and lack of scientific proof that there was a danger (a contention then true for HIV, false for hepatitis). The nonprofit sectors, specifically the American Red Cross, agreed. Annual receipts for the blood products industry were estimated to run as high as a billion dollars. Further, the most organized and articulate political community opposed screening. "Gay groups had already condemned any call for screening of blood donors as 'scapegoating homosexuals.'"[5] Spokespeople for gay organizations demanded that the right to donate blood be nonexclusive. A leader of the American Association of Physicians for Human Rights exulted, "We've preserved not just gay rights but the human right to privacy and individual choice."[6] After the meeting, all of the major organizations came out against screening donors. No action was taken by the public authorities charged with protecting the public's health.

During this period, there were two exceptions to this chorus of "wait and see." First, some physicians and public health officials (e.g., Don Francis of the CDC) courageously called for donor screening. They saw a likely public health danger and felt precautionary measures were imperative before a catastrophe ensued. Those who advocated action were subjected to a variety of attacks; they remained voices in the wilderness as precious months rolled by. Second, a sector of private industry opportunistically identified a potential market niche. "The for-profit blood products manufacturers, however, did not enjoy the cartel on their merchandise that the non-profit blood centers held. With the fear of direct competition for their market, the spokesman for Alpha Therapeutic Corporation announced that his firm, which manufactured Factor VIII, would immediately begin screening donors and exclude all people in high risk groups, including all gays. This position infuriated gay representatives."[7] As the blood banks did not adopt this policy, it had little impact.

An editorial of April 2, 1983, in the highly influential *Lancet* said that "the evidence is not in." Such prestigious words provided sufficient cover for those who defined their interest as best served by not acting (introducing costly measures, causing concern). Even as evidence in favor of the hypothesis of blood contamination accumulated over the next fifteen months—including the identification of the human immunodeficiency virus officially proclaimed in the United States in April 1984 but discovered first by Luc Montagnier's lab in February 1983—the key actors involved all continued to oppose measures aimed at limiting those who could give blood, testing the blood that was given, or testing and treating the blood already in stock: federal agencies (the CDC, the Food and Drug Administration, the secretary for health and human services), private blood banks and their trade organizations, nonprofit blood organizations such as the American Red Cross, and a variety of gay activist groups.

By July 1983, the CDC was convinced of the need for screening. The blood bank industry remained firmly against it. They had powerful friends in the Reagan administration. Reagan officials assured the American people that the blood supply was safe. In July, a license was issued to a pharmaceutical company to start manufacturing heat-treated Factor VIII to reduce the hepatitis threat. However, the treatment was very expensive. Access to it would therefore be limited. The blood industry continued to stonewall, calling for new task forces to study the question. By January 1984, some blood banks instituted a voluntary donor-screening program, but it remained very voluntary and very ineffective. Finally, in February 1984, under threat of a lawsuit, Irwin Memorial Blood Bank in San Francisco broke ranks with the industry and announced a screening program (for hepatitis B) to begin in May. By September, the scientific evidence (and legal pressure) was overwhelming, and nobody any longer debated whether blood transfusions could spread AIDS.[8] This consensus, however, did not result in rapid remedial action.

By January 1985, significant progress on an HIV antibody test was announced. On March 2, 1985, Abbott Laboratories airfreighted the first AIDS antibody test to be publicly released in the United States. It went to Irwin Memorial Blood Bank in San Francisco. Shilts caustically observes that Irwin Memorial, after all, had the dubious distinction of dispensing more HIV-tainted blood than any other blood bank in the country.[9] Hours earlier, Reagan administration officials announced that the Abbott test was to be distributed to 2,300 blood banks and plasma centers throughout the United States. Up until forty-eight hours before the test's release, gay legal groups sought to block its use, fearing donor screening would be used as a tool of discrimination against them. They fought for and won assurances requiring preventive labeling on the kit.

French "Blood"

The hemophiliacs were one of the groups most affected by the status of blood. The early 1980s was a period of optimism for this community. Not only had the life expectancy of a hemophiliac doubled between 1968 and 1979, but quality of life had vastly improved as well—young hemophiliacs were even taking part in competitive sports. The principal reason for this change was the purification and introduction of highly effective bioengineered clotting factors, most famously Factor VIII. Such concentrates were prepared from the blood plasma of multiple donors. Commercial concentrates are made industrially and distributed internationally. Each lot of such concentrates contains the clotting factors from the blood of twenty-five hundred to twenty thousand donors. Consequently, the risk of exposure to a virus was enormous. It was, however, extremely difficult to advocate a return to the pre–Factor VIII methods of multiple injections of blood product from single donors, with their dramatically lower efficacy. Denial may have been understandable, but it was also mortal.

During the 1980s, there were approximately four thousand hemophiliacs in France. By 1982, two-thirds of the concentrates were imported, because the French system could not keep up with demand.[10] The crucial years of HIV infection of hemophiliacs were 1983–84. This situation is even more distressing given that it had been known since the end of 1983 that heating antihemophilic concentrates over 60°C (140°F) for several days inactivated (hepatitis) virus while conserving the clotting power of the material. The American system, driven by profit, dragged its feet. The French system, a state monopoly, encumbered with its own accounting problems and the kind of inertia and careerism so common in a shielded bureaucracy, equally dragged its feet.[11]

France has a national system of blood transfusion coordinated by the Centre National de Transfusion Sanguine (CNTS). The CNTS was created after World War II (1949) and was organized and run according to three sacred guiding principles: benevolence, donor anonymity, absence of profit.[12] Although retaining a certain independence, the CNTS was under the tutelage of the Ministry of Health. The CNTS was in charge of the importation of all blood products. As we have seen, Factor VIII required massive pools of donors and France could not supply these. Hence by the 1980s, despite its principles, France was importing blood. This fact, plus the sheer scale of the organization, meant that by 1984, as its director, Dr. Garretta (one of those indicted and convicted), put it, the CNTS had an industrial enterprise [une structure d'entreprise].[13]

The CNTS did not collect blood, it only treated and distributed it. Collection was in the hands of the Centres de Transfusion Sanguine (CTSs). There were 170 centers dispersed around France; they were of diverse organizational types, from volunteers to patients groups.

In May 1983 an American firm informed Garreta of their heat treatment to inactivate viruses in blood stocks. He did not take any action. Documents released at the trial show without any ambiguity that French authorities knew and believed that the blood supply was dangerous by August 1983. Scientific proof was absolutely established by October 1984. Still, Dr. Jean-Pierre Allain (a respected physician, head of a school for hemophiliac boys, head of anticoagulant research and development at the CNTS, and eventually one of the people indicted and convicted) wrote in the October 1984 issue of the journal *L'hémophile:* "The risk of getting AIDS remains extremely low for a hemophiliac. There is no correlation between the appearance of AIDS and the quantity of blood concentrates injected."[14] The doctor knew better.

On July 24, 1985, the minister of health decreed that untreated stocks were not to be used after August 1 and reimbursement by the state health insurance agency would stop by October 1. A knowledgeable analyst observes: "It would be hard to come up with a better way to stimulate rapid distribution of the old stocks. Beginning in 1985, France imported heated products; even so, several regional centers handed out dangerous lots until October

1985, in some cases even until the beginning of 1986."[15] It was not until May 10, 1985, that the French hemophiliac association voted to stop using stocks of nontreated blood—by October!

Starting in the summer of 1987, parents of affected hemophiliacs began to organize to defend themselves and to seek redress for their exposure to contaminated blood products. Their actions led to a series of lawsuits filed in 1988–91. These led to the trial and eventual conviction of several of the key actors, although many others, including the ministers charged with oversight and their special scientific and medical advisors, were not indicted.

Contradictions

In sum, three things are clear about the French situation: (1) those who gave blood could have been better screened; (2) known contaminated stocks could have been treated or discarded; (3) the introduction of the HIV antibody test was delayed. The status of blood in France was anything but "transparent." Although blood was a merchandise that was nominally *hors du commerce,* there were massive costs in collecting, storing, treating, and distributing blood and its subproducts. Different steps in this process were subject to different economies. For cultural and political reasons, it was basically impossible to say outright in France that the French system was a commercial one. But, ultimately, it was.

The CNTS did not want to simply dispose of the stocks of blood they had stored. They had the legal right to dispose of them. However, they were under commercial and bureaucratic pressure not to do so. They didn't. Further, the French authorities wanted to produce heated concentrates themselves. They did not want to use "foreign" tests. The estimates of the number of hemophiliacs infected by this decision varies between 10 and 1,000.

The first commercially available antibody test for HIV (there was no DNA test at that time) was "American." French officials did not want to employ American tests or procedures, mainly for financial reasons, but also for nationalistic ones. The Pasteur Institute was developing its own antibody test but it was not yet commercially available. On February 11, 1985, the makers of the Abbott test formally applied for a license in France. The test did not receive authorization even though it had been approved in the United States, Germany, and Australia. French authorities demanded more proof of efficacy (the test did have a slight false-positive rate). Notes released from a May 9 meeting of the prime minister's cabinet spelled out their concern: "The moment the tests are authorized the French market will be largely captured by the American test. . . . [Therefore,] the cabinet of the prime minister requests . . . that the Abbott registration dossier be retained for some time by the National Public Health Laboratory."[16] The Abbott test was not finally approved until June 1985 and was not put into wide circulation in France until August.

What prevented France from screening donors? The answer lies in the idea that the gift of blood is free, generous, and benevolent. It was impossible to question the validity of the key symbolic component of the system. To do so would have been to challenge the most "sacred" values in French secular society.[17] It was not done.

Benevolence

In France, the collection of blood falls into the domain of the gift, of benevolent gestures, of solidarity. This symbolically charged realm has been given a distinctively French form. Although donors were not "paid," they were "reimbursed." Such transfer of money (for time lost, for travel, for a hearty lunch) was carefully encased in symbols of precapitalist exchange or noble gestures.[18]

France, of course, was not alone in holding this position although its specific form and symbolism were distinctive. In the English-speaking world the classic presentation of the case for altruistic donation of blood is Richard M. Titmuss's *The Gift Relationship: From Human Blood to Social Policy* (1970), which stands unchallenged even today as the best statement of the case for a donor system. Titmuss concluded his study in ringing terms: "From our study of the private market of blood in the United States, we have concluded that the commercialization of blood and donor relationships represses the expression of altruism, erodes the sense of community, lowers scientific standards, limits both personal and professional freedoms, sanctions the making of profit in hospitals and clinical laboratories, legalizes hostility between doctor and patient, subjects critical areas of medicine to the laws of the marketplace, places immense social costs on those least able to bear them, increases the danger of unethical behavior in various sectors of medical science and practice, and results in situations in which proportionately more and more blood is supplied by the poor." The altruistic system was also more economically and administratively efficient, and "in terms of quality, commercial markets are much more likely to distribute contaminated blood; the risks for the patient of disease and death are substantially greater. Freedom from disability is inseparable from altruism."[19] Although triumphant in 1970, the next two decades saw a certain erosion of the system upon which these moral premises rested.

The newly expanded 1997 edition of the book contains a number of new chapters that analyze a changing situation. Virginia Berridge, in a chapter titled "AIDS and the Gift Relationship in the UK," reminds her readers of the consensus in 1970 on Titmuss's position, most generally around his claim that "transfusion systems were markers of the morality of the society which produced them." However, by the 1990s, in England, the "commercial/bad, voluntary/good dichotomy helped determine the direction of policy within the Blood Transfusion Service, but in reality, other factors, including the impact of technological change, made the starkness of the opposition less relevant. One of the striking features of the blood system highlighted by AIDS in the UK, in France, and elsewhere in Europe was that a 'volunteer image' fronted systems which had become highly dependent on commercial systems. Whereas the blood donation side of the transfusion service dealing in 'whole blood' was volunteered-based in the classic way, blood products were primarily commercial in origin, part of a huge international trade in blood."[20] This disparity between moral claim and institutional base did not turn into a crisis in England because the Thatcher regime made cost-accounting procedures one of its major reforms of the National Health Service and consequently there was extended public debate about what services cost. Finally, technological procedures concerning heating blood products were introduced in Britain earlier and not under a cloud of crisis and controversy.

In France, almost from the beginning of the system, however, and ever more so after the introduction of Factor VIII, production could not meet demand. Hence, foreigners and commerce could not be avoided, and foreign blood was imported. Efforts continued to shore up the French system, to protect its principles, and to increase national production. In the early 1980s, the Socialist government expanded its campaign to collect blood in French prisons. In retrospect, this decision was disastrous. Here was a humanitarian gesture reminiscent of the eighteenth century, a means for prisoners to alleviate their social debt, to achieve social "inclusion." Social scientists supported the program of prison donations for its humanitarian inclusion; prisoners apparently enjoyed the break, as well as the wine and sandwiches, and blood bankers appreciated the blood "they could collect during traditional lean times of holidays and vacations."[21] Regrettably, observers agree, this program was in all likelihood the greatest source of contamination of the French blood supply.[22] Later, after the scandal, during a parliamentary debate on a law (January 4, 1993) modernizing measures for ensuring the safety of blood transfusions, Bernard Kouchner, a founder of the humanitarian Doctors without Borders (Médecins sans Frontières) and the minister of solidarity testified that mistakes had been made, for good reasons. It was not an easy task, he testified with a certain defiance, to "take charge of this wonderful effort of solidarity of our two million benevolent donors who constitute our patrimony and demonstrate our generosity."[23] It was simply unthinkable to put this part of the French system in question. The affair of contaminated blood would not be about the unthinkable.

Marie-Angele Hermitte, in her book *Le Sang et le droit: Essai sur la transfusion sanguine,* proposes to elucidate "the mystery of benevolence and nonprofit."[24] She comments that writing the book proved to be a particularly difficult and soul-searching task. As a legal specialist as well as a staunch defender and theorist of bioethics, what she uncovered was not comfort-making. Her book outlines in great detail the history of transfusion and its current legal status in France and the European Community, a community sharply and often schizoidly divided between the Europe of Commerce, centered in Brussels, and the Europe of Human Rights, in Strasbourg. It was not unknown for French delegations in the different cities to defend radically different conceptions of the line between commerce and human rights.

The story of successful blood transfusion is a twentieth-century story. Blood groups were discovered in 1901. It is also a story of mass warfare and the consequent need for plenteous supplies of blood. Seen from the present, the decisive moment in the French history of blood transfusion took place during World War II. In order to supply the French Resistance with the blood it needed, a system of volunteers, of *bénévoles,* was organized. From this patchwork of individual acts and organized collections was forged a powerful and enduring symbolism, a noble ethos—of the act of giving blood as the embodiment of resistance. After the war, for some this act symbolized resistance to capitalism; for others, resistance to evil; for others, to the materialism of the modern world.

The actual events are instructive. The field of experimentation of mass collection of blood based on benevolence and solidarity was North Africa, where the eventual invasion of Nazi-held Europe was being prepared and where Free French Forces were operating. A popular campaign to collect blood was most successful among the Berbers of the Kabyle region of Algeria. North Africa was also the site of other technological innovations, such as the first industrial processes of blood dessication. As fascinating as this theme of the colonies as "sites of experimentation for modernity" is, it is not the theme of this book. More importantly, and not surprisingly, these mass efforts and technological innovations did not become the focus of French symbolism. Rather, the substance of symbolic elaboration was the heroic acts of individuals in the French Resistance: brave citizens and soldiers offering the gift of life to their comrades defending France. The contrast with the policies of the collaborationist Vichy government could not have been more charged. Blood donors were given a pass, an *Auswies,* allowing them extra food, telephones, gas rations, release from certain guard duties, fewer travel restrictions, and other such valued rewards. It was not hard to make the equation "benevolent donors = resisters, paid donors = collaborators." After the war, transfusion carried with it the mark of "solidarity, of a voluntary and benevolent gesture, of a collective effort of the entire nation. Benevolence had become the mark of France's French liberators."[25]

The system—or at least the symbolism—became a spontaneous, spectacular, and enduring success. Gradually, the "other" of this blood donation system became the United States, which replaced the Germans and Vichy. The French system is a magnificent one of social solidarity based on an act, the passing of the most symbolically precious of substances, thereby creating a tie, a connection, a bond, a recognition. The *bien* (good), by being given, creates a *lien* (tie). Once established and institutionalized, the system attained truly extravagant legitimacy; a survey in 1992 showed that 96% of French people favored it. The system not only created a social bond but worked the miracle of creating and providing the "warrant for a system of industrialization without profit."[26] The twentieth century has seen a proliferation of symbolisms of industrialization without profit. However, right-wing regimes, left-wing regimes, antimodern revolts, and postmodern discourses have so far produced a meager harvest of stable institutions capable of bringing this oxymoronic site into being. Profit, we now know, takes many forms, and production and industry are not always happily joined. Nonetheless, it must be admitted that the French blood system is no doubt one of the most successful and admirable attempts to bring this hybrid into existence and to sustain it through thick and thin.

Hermitte proposes a startling and troubling conclusion. An unexpected price has had to be paid. During "all this period of ideological warfare, it has never been a question of the superior quality of donated blood. The benevolence system posed the question of human dignity without connecting it to that of public health. This argument was only added later. We know it played a very negative role after the contamination by the AIDS virus."[27] This disproportion between the public good and public health is troubling. In the French distribution of things, there has never been any question that protecting the donor is the cornerstone of the system. The primary social link is between donors. The link with those who receive the blood is secondary.[28] A principle of the French system is the anonymity of the donor. Such anonymity applies once the blood is given; the blood disappears into a distribution system, one that was, and needed to be, a bureaucratic and economic institution, albeit of a special kind.

Commerce

In France, "blood" is a "thing" that is juridically *hors du commerce,* something that can be neither freely sold nor exchanged. When such a thing is an abstract possession (e.g., nationality or the right to vote), no problem is posed to understanding what is meant and to whom it applies. The situation becomes more complex when this category is applied to material things. The concept of "patrimony" is the best example of material things that can be bought and sold but are still regulated in such a manner that their value is not solely determined by the market. The matter becomes more complex yet when the substance in question is an element of a living system. It is not that a human cell has a special status in and of itself; rather, things like cells take on a special status only because, Hermitte writes, they concretely imply "the person who is the source and, in a more abstract manner, the representation that is made of humanity."[29] Hermitte herself had previously attempted to articulate a philosophical and legal doctrine that would define this specialness, this "dignity" of *la personne humaine* and *l'humanité*. We will return to this issue at more length later. Regardless of precisely (or imprecisely) what the terms mean or signify, there is a strong tension between public health and the "dignity" of the human person as embodied in things. At certain times, under specific circumstances, these two logics of value, of dignity, and of solidarity can find themselves, to return to our vocabulary of purgatory, heteronomically associated. Public health operates under the economy of "good administration"; one needs to know what resources are available and to take them up as things to be collected, stored, and distributed as efficiently as possible. The economy of the *bénévole* system differed: "the 'plus,' its symbolic difference, implied by the benevolent system is respect for human dignity."[30] The blood transfusion issue was one moment when the two logics of value clashed.

Finally, it is worth noting one additional aspect of French "blood." Blood contains albumin. So, too, does the placenta. Albumin changes legal status in France depending on whether it is taken from blood or from the placenta. In French law, the placenta is *un déchet* [a waste product]. Consequently, it can enter into the commercial realm, and it does. Albumin is routinely extracted from the placenta and sold. The symbolic charge is very low; there is no public outcry about dignity or the rights of women. Still, French legalists have bent over backward to justify this (noncontroversial) status of a body part sold without explicit consent. Even though the mother may never have authorized (or known about) the sale of the placenta, she is presumed to have given her consent—the legal scholars have been ingenious here—because a mother has an implicit duty on behalf of her child to "maximize the patrimony" [faire rendre le maximum au patrimoine].[31] Presumably because France has a state health care system, the proceeds from the sale (to pharmaceutical and cosmetic companies) of the albumin-rich placenta return, indirectly to be sure, to the nation's patrimony and hence to the child's inheritance. Is albumin a thing to be bought and sold? The answer is "yes," when it is extracted from the placenta and "no" when its source is blood. Another legal scholar, Jean-Pierre Baud ironizes, "nothing demonstrates better that our law governing human body parts is closer to savage sacralities than to modern physiology."[32] Baud is a secular modern and a rather militantly traditional one at that.

So, too, although a little less militantly and with less assurance of where she stands, is Hermitte. When it comes to sacred substances, she observes, with a certain regret, that in such matters there are national histories, even nationalist ones. "Each people has found its own solutions, profoundly anchored in its past."[33] The past, of course, is a collective construction, especially when it comes to a sacred one. That in no way diminishes its vitality, but only makes more transparent what form it takes under modern circumstances—a social patrimony.

Comparative Contradictions

In both the American and the French system, a clear conflict of interest existed between public health and the institutions charged with dealing with the blood supply. In the American case, there were very large and very well represented financial interests at stake. These interests had a sympathetic, pro-"free-enterprise" government who listened to them. Denial and bureaucratic inertia prevailed. There was also a "human rights" dimension to the events. Specifically, a well-organized and vocal gay community insisted that any form of restriction on who could give blood was a potential assault on human rights. They vigorously opposed donor screening until the very end. This combination of business interests and human rights considerations put heavy pressure on the public health authorities to be very cautious, and thus, they were scandalously slow in taking action.

In the French case, commercial interests were involved as well, but they were more hidden. They were not as large in a quantitative sense, but because of how they came into play, they were very important. The CNTS was a business. It was a state-run business but a business nonetheless. It had its financial and national interests to protect. It delayed the entry of foreign tests and foreign technologies that would have cleaned up the French blood supply. The CNTS also refused to destroy stocks of contaminated blood when it should have. The CNTS was under the direct tutelage of the French (Socialist) government's Ministry of Health. The Pasteur Institute lobbied the French government successfully to delay the American antibody test for HIV. They had their own test in the works and did not want to lose their home market. They had powerful friends in the right places. However, as we have just seen, there was another dimension to the French situation. Whereas in the United States the "principled" opposition to screening donors was led by the gay community in the name of human rights, especially nondiscrimination, in France principled opposition was embodied in the donors' associations. Their contribution cannot be underestimated. It provided momentum to a policy of not screening donors on the one hand and to a certain nationalism on the other. This subjectivism (both individual and collective), defended with the best intentions, had disastrous results.

The *bénévole* system of solidarity remains paradigmatic in France. Its tragic interference with bureaucratic and financial economies in the contaminated-blood scandal discredited, not the icon, but only a few of those charged with its stewardship. A human supplement, a social discourse, and a political will give this system a distinctive form. The system was within the normativity of a *bios*. The proverbial heart of the system was the donors. The quality of the blood they gave was secondary to the manner in which they gave it. Ultimately, the state of the blood itself was a technical issue (however mishandled it may have been). Jean Dausset had extended the *bénévole* system to HLA typing and carried its networks beyond the state although he was successful only at the level of scientific exchange (e.g., he was unable to establish an international network of organ exchange).

If the blood donor system was the epitome of *bios* in the postwar French articulation of biopower, then the rise of genomics remains exterior to, and poses a dramatic challenge to, such a *bios.* Obviously, this is not to say that genomics is external to biopolitical assemblages or that attempts such as those of the AFM to give a new *zoē* a new form failed. But it is too soon to tell what form or forms will emerge and be stabilized for an extended period of time. In the meantime, the situation remains dynamic. The AFM has provided one initial and impressive model of how to bring genomics and solidarity-producing sociality together. Whether the "les malades," or "patient," model can be or should be generalized remains to be seen. The question remains open: What is the social form (or forms) to be given to molecular information, sequencing machines, gels, data banks, tissue collections, and patients groups' avid cooperation? In what way is this realm of hundreds of thousands of base pairs, of sequence-tagged sites, and of arbitrary genetic markers the "same" as the iconic acts of solidarity formed during the Resistance? If there ever was an example of brute life, the minimalist essence of beings, it is genomics. The blood taken from donors for transfusion is still a sacred substance, life-giving to another. Genomic information has none of this archaic symbolism, even though efforts are certainly being made to invest it with a profound and pervasive aura of spirituality. What form could be given to this state of *zoē* that would transform it within a transnational realm, without any articulated sovereignty, whose driving forces are money, technical and scientific research, biosocial interest groups, and ethics committees?

French Bioethics: Alarm and Repent

During the 1970s and 1980s in France, there was a substantial demand for popular books on biological and medical themes. As the following list indicates, this demand was met. The form in which that demand was met (and, to an extent, created) is distinguished by what might be called its style of "high vulgarization," the prominence of writings by representatives of the mandarinate of French medicine and science. The author of an excellent study of the French National Ethics Committee observes:

> With the appearance in 1970 of two books by Nobel Prize–winning biologists, *Le hasard et la nécessite* by Jacques Monod and *La logique du vivant* by François Jacob, there was an avalanche of books about medicine, its dramas and practices: *La puissance et la fragilité,* by Jean Hamburger (1972), *Grandeurs et tentations de la médecine,* by Jean Bernard (1973), *Le mandarin aux pieds nus,* by Alexandre Minkowski (1975), *L'homme changé par l'homme,* by Jean Bernard (1976), *L'homme et les hommes,* by J. Hamburger (1976), *Changer la mort,* by Léon Schwarzenberg (1977), *L'espérance ou le nouvel état de la médecine,* by Jean Bernard (1978), *Demain les autres,* by J. Hamburger (1979), *Médecin de la liberté,* by P. Milliez (1980), *Mon beau navire,* by J. Bernard (1980), and *Un Juif pas très catholique,* by A. Minkowski (1980).[34]

These books, the media, and proclamations of the National Ethics Committee, as well as other currents of opinion and opinion makers, created a rich and internally contradictory imaginary, one fulfilling de Certeau's criterion of a *lieu pratiqué.* The world portrayed was both ominous and reassuring. One dominant trope presented images and narratives of Progress, in understanding, in control, in therapy; another, contrastive one presented a variety of brave new worlds peopled with malign offspring, the product of barbarous human forces (often) carried by foreigners.

In 1978, to great publicity, British doctors announced the birth of the first "test-tube baby." This event resonated like a great tolling bell in France, acclerating and focusing reflections and reactions, multiplying calls for action, for clarification. Condemnation was rife among those with a right to be quoted; a demand existed, however, in society that could be silenced but not stifled. Authorities were troubled, hesitant, agitated. It is plausible to point to the reactions to this technical procedure during 1978–79 as a moment of a major reformulation of French bioethics: Paris hosted the

first major colloquium on bioethics (dealing with artificial insemi-
nation); it authorized the first university and CNRS research
teams.[35] During the 1980s, procreative issues were the main terrain
of bioethics in France. The hard-won and belated right to abortion
was not the center of the debate in France as it arguably was in
the United States. Like so many profoundly controversial issues
in France (Vichy, the Algerian war), abortion remained in the
background. It was cast as one element in the citizen's right to
a "free disposition of the body," even if, typically, its regulation
remained in the hands of the state and was never completely re-
linquished to the woman. With a tacit concordat on abortion,
artificially assisted procreation took center stage. In 1982, French
doctors announced the first French success at in vitro fertilization—this announcement precipitated the formation of the National Ethics Committee in 1983.

Jacques Testart's description of the artificial procreation technique, in whose invention he played a major role, captures the
mood extremely well while contributing to its construction and
dissemination:

> Take a close look at the instruments thanks to which one
> assists the procreation of modern man. First the receptacle
> where the seed oozes, next a cylinder the size of a thumb
> and as long as a hand, whose ample antechamber is ringed
> by a vulval fringe. The chalice for the virile offering—
> is it a negative phallic shape or a vaginal mold? Then
> comes the tube where the gametes are married. The tube
> is frail and long, with detours. . . . The instruments have
> something in common, they have a thin and transparent
> surface. The receptacle, the tubes, the incubator, and the
> catheter are only conventional limits, designed around
> various cavities, cavities that one also calls "enlightened."
> The seeds of the human male and female today voyage
> through these vessels of light. The egg is transparent. . . .
> What remains of fantasy when it is taken over by tools?[36]

Testart publicly imposed a moratorium on his own research
in 1985. Grandiosely repentant and compellingly eloquent, Testart embarked upon a highly visible (well-orchestrated and well-

received) media campaign warning in apocalyptic terms against the potential threats to humanity posed by his own previous bio-medical practice. In his best-selling book *L'oeuf transparent* (The transparent ovum), with a preface by the philosopher Michel Ser-res, Testart nailed his theses to the proverbial door: "I claim the right to a logic of nondiscovery; an ethics of nonresearch. Let us stop pretending that research is neutral and that only its applica-tions are qualifiable as good or bad. Let someone demonstrate a single example of a discovery that was not applied when it ad-dressed a preexistent need or one created by itself. It is before the discovery that one must make ethical choices."[37] Testart speaks of "research" in the singular. He speaks of a simple and ineluctable logic of history. He speaks of choices to be made, stands to be taken. He speaks of unknown forces, of disembodied needs, of great dangers. He speaks of "ethics," not politics. Testart's act in-troduced an imperative of penitence into France's emergent bio-ethical discourse. Perhaps more accurately, it triggered the recog-nition of a modern French bioethical purgatory. What the "silent scream" video was to the antiabortion movement in the United States, Testart's book was to bioethics in France. The difference of course was that the U.S. movement was self-identified as reli-gious and right wing, while in France the alarmed, aroused, and spiritualized humanitarianism either was situated on the left (Testart was a former Trotskyite) or, as with the National Eth-ics Committee, sought to place itself above partisan politics alto-gether.

Ethnographically, the twinned themes of militant penitence and vigilance are distinctively resonant in the French case during the 1980s. However, not surprisingly, so too are proclamations of hope. For many in the biosciences—and especially in certain pa-tient groups—new weapons were becoming available to overcome suffering with which humanity had too long been afflicted. Ap-propriate intercessions were no doubt required for humanity to progress in the right direction. But what form should those inter-cessions take? Response ranged from calls for "moratoria" to a "genomic Manhattan Project." All agreed, however, that some-thing extremely formidable for the future of humanity was at stake. To understand how words and things took on this distinc-tive cast in France, some comparative background will be helpful.

Bodies and Institutions

In the name of dignity of the person, French law basically refuses the individual the right to dispose of his or her body and its parts; American law has allowed greater latitude for proprietary and commercial relations concerning the body and person, privileging autonomy and value over an inherent and inalienable dignity. Autonomy contra dignity, dignity contra autonomy, antinomies linked in an uneasy seesaw, with neither tradition totally eliminating what the other valorizes: ultimately both positions reluctantly return to an uneasy and unstable privileging of the human person—and hence a spirituality that is hard to stabilize. In both cases that person is no longer uniquely Kant's autonomous, rational being but rather a living being, who is somehow something, somebody, someone, very special. Many of the current bioethical controversies, as well as the political and social struggles in the bioethics arenas, refer either to the unborn or to those at the end stages of life whose mental capacities are severely limited, if they still exist at all. Battles over abortion, euthanasia, and new technologically mediated definitions of life and death create issues that science and technology help to produce but that, as Max Weber so clearly enunciated in his 1917 "Science as a Vocation," they have no power to resolve. Yet, the demands to do so remain vital. It is safe to say that neither scientists, feeling compelled by the quest for funds to make predictions of imminent breakthroughs, nor humanists, exempt from legislative powers yet eager to justify their own enterprise to the public, have always been able to measure up to the rigorous asceticism Weber demanded of them.

The case of *John Moore v. The Regents of the University of California* encapsulates many of the key elements in contemporary American debates about the body, what its boundaries are, whether its parts can be owned, whether its parts embody the person. John Moore sued the University of California after the doctors at UCLA Medical Center used cellular matter removed from his body—in the process of successfully treating his cancerous spleen—to produce an immortal cell line which they then patented. Moore claimed a share of the profits, arguing that the cells were his property. The California Supreme Court ultimately did not agree with Moore. The court ruled that Moore did not have "ownership or right of possession" (which a French court would agree with). They did rule that the doctors had violated their ethical and legal responsibility in not telling Moore what they had done with his cells. Even though the judges did make a ruling in the legal case, they were hesitant about the implications of their ruling. A conservative justice wrote: "Plaintiff has asked us to recognize and enforce a right to sell one's own body tissue for profit. He entreats us to regard the human vessel—the single most venerated and protected subject in any civilized society—as equal with the basest commercial commodity. He urges us to commingle the sacred with the profane. He asks much." The doctors, however, were not so constrained, even if the judge felt the state had a role to play in protecting the public's interest. A progressive justice, who opposed the majority decision, wrote: "If this science has become science for profit, then we fail to see any justification for excluding the patient from participation in those profits. . . . Our society acknowledges a profound ethical imperative to respect the human body as the physical and temporal expression of the unique human persona." Commodification violates "the dignity and sanctity with which we regard the human whole, body as well as mind and soul."[38] These basically religious appeals to "the sacred" and "sanctity" underline a malaise over a major boundary shift but hardly constitute a coherent doctrine. Durkheim's invocation of the sacred as the ultimate symbol of society seems more apt, if, as we shall see, equally vague and contestable.

The distinction between the person and the body is further complicated in current debates by the unstable relationships of body and things. Jean-Pierre Baud in his *L'affaire de la main volée: Une histoire juridique du corps* underscores the ponderous heritage of the division between *persons* and *things* fundamental to Roman law. In Roman law, the person is a fiction that identifies the individual as *une personne morale* on the judicial scene. The moral person is the holder of rights. In 1810, when the French code was promulgated, keeping the physical person separate from the moral-legal *persona* did not pose a problem. In line with Roman law, those things that could be physically separated from the body fell neatly into two categories, both exempt from civil law: "nuisances or burials" [des nuisances ou des sépultures]. The codes of hygiene (and public health) and mortuary practices of religion (or the Republic) took care of these polluted and sacred things. Today, the situation is not so simple. Separable, exchangeable and reincorporable body parts (or parts of parts) confuse and confound the previous legal framework.

The investment modern society has made in technological change has produced results (provisional answers and truths) and these have occasioned further social demands. Frequently, it is not clear in advance which ones are taken to be problematic. Thus, for example, the first text in French law permitting a gift of organs, the eyes, a decree of July 7, 1949, proved to be uncontroversial. However, new techniques for storing blood made transfusions on a mass scale possible and became the subject of an active debate. They soon became a social fact to be reckoned with. The French law of July 21, 1952, on transfusion became the first legal fissure in the corpus allowing "living" bodily parts (neither excremental or alimentary) to be transferred. This time a technological advance was taken up as a moral and social problem. "For the first time since the installation of Roman law, it became necessary to question the status of something human, living, but of which one could no longer say that it was constitutive of the person."[39] From the debates and reflections that ensued arose the principle of the freely given gift [gratuité du don]. The principle of nonremuneration for such gifts has become the cornerstone of French bioethics. In the light of this principle, the situation remained stable for some three decades. For example, there was little debate surrounding the law of December 22, 1976, on organ donation that made organs medically available unless the living person had explicitly stated the opposite. Taking organ transplants as equivalent to blood transfusions, the organ law was seen as facilitating a public utility. However, with the rise of biotechnology and the specter of genetic manipulation, as well as the internationalization of the problem, the debate over the body (and its parts) as thing and/or person was reawakened during the 1980s.

The stage for bioethical developments was set by a series of laws dealing with life's major thresholds (birth, death) and subsequently with a rethinking of the French position on the circulation of body products. It is not much of an exaggeration to say that everything passes through the state in France. Nor is it an exaggeration to say that while each piece of legislation dealing with these major issues is solemnly promulgated in the name of grand principles, the whole legal corpus is not much more coherent than the American one. For example, although prostitution was legal in France (it still is), contraception was not until 1967. The 1976 law on organ donation assumed consent for harvesting except if there is an explicit statement to the contrary (the presumption for the cornea, however, is the opposite). Hair, nails, teeth, and (until recently) milk are salable in the marketplace. It was standard practice to sell placentas and umbilical cords, which were used by the cosmetics and pharmaceutical industries. Such matter was considered to be technically "abandoned" by French law and hence required neither a counterpayment nor a consent form. The French legislation legalizing abortion in 1975, 1979, and 1982, while unquestionably a major change, clearly states that abortion is not an exlusively private matter of a woman's choice. The state both formally recognized a changed social reality and retained its role as regulator. A parallel situation obtains with the 1988 legislation on death and terminal illness. A 1993 law allows a margin of commercialization of blood products as long as they have been collected in the classic altruistic fashion (*gratuité, anonymat, volontariat, respect des conditions sanitaires*). One commentator concludes that these vacillations leave "the impression of a major hesitation between extreme positions that brings about a set of policies that are contradictory, changeable, and heterogeneous."[40] There is simultaneously a recognition of the need for change, a sustained suspicion of the market, and a recourse to the state.

One significant factor contributing to this changed situation has been the dramatic changes in the life sciences, especially molecular biology. The institutional and the scientific history of molecular biological research in France since World War II have been closely intertwined. The institutional umbrella under which change was to take place was the Direction Générale de la Recherche Scientifique et Technique (DGRST), charged by President de Gaulle in 1958 with implementing "coordinated programs" in order to fill scientific gaps in applied fields. Responding to a widespread consensus on the need to modernize and reorganize French scientific and technological institutions, a distinguished group of experts made a series of recommendations for reorganization. Nobel Prize winner Jacques Monod headed this subcommission. They identified molecular biology as the most important sector. Between 1960 and 1970, new CNRS institutes were established; funding was made available to extend the use of molecular biology methods and concepts to immunology, embryology, and physiology. The government policy favored small, interdisciplinary problem-oriented units rather than large centers. Although this strategy served French science well for almost two decades, it proved an initial handicap when it came to genomics, which operates at an industrial scale, requiring large investments for machines and labor.

Comparatively speaking, the site of bioethics debate in France is distinctive in its centralized, state-supported setting, its linking of scientific advance with human rights, and its historical cast. The DGRST included an ethics component as a direct response to the 1974 Asimolar conference on recombinant DNA. Although its initial task was restricted to classifying genetic engineering facilities in France, it nonetheless established a principle and an institutional precedent. In 1983, President François Mitterrand created "un Comité consultatif national d'éthique pour les sciences de la vie et de la santé," the first national ethics committee. It was charged with setting out ethical guidelines for research. The committee's mandate is to inform the public and the government of its reflections on the long-range implications of medical and scientific advance. It holds an annual day of public debates on relevant bioethical issues, regularly publishes documents on current issues, etc. This mandate was enlarged during the course of the decade.[41]

The national committee sought to be both a mirror of society and, in Republican fashion, to be above society. Offshoot committees formed to deal with specific crises and reforms were named "Committees of Sages." The committees' role then was both to bring issues to the public's attention and to provide a forum in which those issues could be clarified and general principles defined. Although the members of the commission were chosen to be representative of French society and although there was some change of personnel during the first decade of the commission's existence, there was a remarkable stability; the core participants numbered in the vicinity of one hundred and fifty in ten years. The smallness of this number indicates the contours of the elite from which they were drawn. These were, apart from a smattering of religious and confessional representatives, members of the liberal professions, especially law and medicine. There were scientists involved, especially some notable molecular biologists such as the Nobel Prize winner François Jacob, but they tended not to be active in formulating positions.

In a report titled *Sciences de la vie: De l'éthique au droit,* written and delivered in 1988 in response to a formal request from the French president to the Conseil d'Etat, the French position was officially clarified. As Jean-Pierre Baud puts it, French doctrine "expresses itself in the simplicity of an axiom and with the ambition of a mission: the body is the person; this is one of the modern elements of France's eternal civilizing mission—to make this principle triumph over the mercantilism of industrial society."[42] The body is a thing but one that is not allowed to circulate in the world of commerce: "Human dignity forbids that man be given a right to own his own body." In line with the divisions of Roman law, the body falls either to public health or to the sacred. The body does not belong to the person but to French society. By the mid-1990s, legislation to this effect was proposed and passed by Parliament.

French reflections on "human rights" provide both a statement of universal moral principles and a means of translating those principles into legal and quasi-legal administrative procedures. A contrastive universalistic set of moral claims, also tightly linked to legal and administrative concerns, exists in the United States. Remarkably, at least in hindsight, what came to be known as bioethical principles and practices were not directly based on the judgments handed down by the Nuremberg tribunals and formulated as the Nuremberg Code. David Rothman in *Strangers at the Bedside: A History of How Law and Bioethics Transformed Medical Decision Making* documents the scant journalistic coverage given to the work of the Nuremberg tribunal in the United States. He also documents how most of the principles articulated in the code were eventually adopted in the United States but only decades later, after a series of specific scandals—and the changing context of American politics and culture—made them visible one by one. Principles such as "informed consent of the human subject" to experimentation or treatment or that the person be mentally competent to make an "informed decision" were long delayed in the United States.[43] The abuses of Nazi science and medicine were comfortably distanced by attributing them to unique pathologies of Germany and to poor-quality science and medicine. The main lesson drawn from the Nazi horror was that the state should not regulate medicine.

A series of well-documented and well-publicized abuses of medical and scientific experimentation in the United States was not sufficient to change this attitude. It was only during the 1960s, when trust in the government and those previously relatively sanctified authority figures began to erode due to opposition to the Vietnam War, the Civil Rights movement, and feminist politics, that scrutiny of scientific and medical practices began to gain political ground and produce normative institutional changes. Protection for human subjects was introduced first in laboratory settings governed by the ever-increasing power and funding of the National Institutes of Health (NIH). Funding for medical research at the NIH had gone from $700,000 in 1945, to $36 million in 1955, to $436 million in 1965, to $1.5 billion in 1970.[44] By the mid-1960s, pressure to introduce some codified scrutiny of experimentation on humans mounted and bore fruit with the creation of NIH regulations, importantly augmented by similar regulations on consent introduced by the powerful Food and Drug Administration.

During the 1960s, public debate took shape around how to develop an equitable policy for the use of improved (but scarce and expensive) technology for kidney dialysis. Improvements in transplant success rates were cast in similar terms. By the early 1970s, the tremendous publicity given to heart transplants (Christiaan Barnard performed the first transplant in South Africa in 1968), and the evidence of how unsuccessful they were, produced national hearings in Washington and eventually federal commissions. Arrogant testimony by Barnard, defending absolute autonomy for doctors and threatening that if any controls were imposed, American medical research and practice would lose out in international competition, helped push a liberal Senate to take further steps. Similarly arrogant testimony by Nobel Prize winner biochemist Arthur Kornberg of Stanford further impelled the Senate to action. When asked about a scientist's responsibility to the public, Kornberg declared: "There are absolutely no scientific rewards, no enlargement of scientific skills that accrue from involvement in public issues. The biochemist who deals with molecules cannot afford any time away from them. If the research worker were to become a public figure, it would destroy him as a scientist."[45] These attitudes spurred congressional action, which was temporarily blocked by the Nixon administration but eventually succeeded after his departure.

There were two key events in 1973. First, the Supreme Court ruling *Roe v. Wade,* granting the right to abortion as a privacy right. The second was less noticed at the time but gained importance in following decades. Section 504 of the Vocational Rehabilitation Act included "handicapped" people under antidiscrimination law, thereby extending this principle to an ever-greater segment of the American population. An act of the American Congress in July 1974 founded a National Commission for the Protection of Human Subjects of Biomedical and Behavioral Research. It issued the influential *Report on the Fetus* a year later, which resulted in federal regulations; the standards it proposed for the protection of human subjects gradually were accepted as world standards. The commission met for four years and issued ten reports articulating the principles of "respect for persons, beneficence and justice." Above all else, it placed great stress on patient autonomy. Many important issues were delegated to its successor, the President's Commission for the Study of Ethical Problems in Medicine and Biomedical and Behavioral Research, which stood from 1980 to 1983. Its founding documents are *The Ethical and Legal Problems in Medicine and Biomedical and Behavioral Research* (1983). One in particular was significantly titled *Defining Death.* The presidential commission also raised the issues of genetic screening and human gene therapy. It articulated the principles of beneficence, autonomy, and fairness. It was around these principles that the influential American bioethics profession rose to prominence.

This trend of advocating patients' rights and autonomy was not without its critics (aside from those within the research and medical worlds who resented restriction on their autonomy). Renée Fox, a leading medical sociologist and pioneer in analyzing the social dimensions of transplantation, cautioned that "in the prevailing ethos of bioethics, the value of individualism is defined in such a way, and emphasized to such a degree, that it is virtually severed from social and religious values concerning relationships between individuals; their responsibilities, commitments, and emotional bonds to one another. Claims to individual rights phrased in terms of moral entitlements tend to expand and to beget additional claims to still further individual rights."[46] Autonomy in the U.S. context of the 1980s and 1990s had become, ironically enough, entitlement, protected and assured by the state.

Human Dignity in Danger

During the second half of the 1980s and the early 1990s, issues of genetic engineering joined those of artificial procreation as sites of danger and anxiety. Discursively it was held that the body was being parceled into smaller and smaller parts, "alienated" in the eyes of some from the organism as a whole, first organs and fluids and then, the ultimate unit, the "gene." Furthermore, scientists were subjecting these separable things to manipulation and alteration. While these transformations and transmogrifications were still of limited scope, it was widely repeated by foes and advocates alike that they offered the possibility, presented as an inevitability, of a nearly unlimited potential in the future. Representationally, the body was becoming simplified, treated as sheer raw material. In France, the complex affects that these events and their representations aroused turned less on a narrative of "denaturalization" in the American sense (ecology was a minor current) than on a frightened and angry reaction against the demystification of the largely unexamined equivalence of the body to the person. The artificial procreation techniques put into question tacit cultural understandings of kinship relations as well as the associated links between patrimony and inheritance—identity and property were at stake. For a (post-) Christian, (post-) bourgeois society, the threat could not be more ominous. The tacit identification of the body and the person, and this was to a large extent the cause of the shock, had not previously been consciously scrutinized. The body was taken to be sacred, holistic, the container of the past and the vehicle of the future. In France, the sacred, since Durkheim, had been society. Its fate was now uncertain. Body parts were entering into a machinery that produced spiritual entities.

The French National Ethics Committee frequently exhibited

a malaise that was formulated in terms of a defense of the integrity of "humanity" and "the human person." It was troubling to the guardians of the French Republic that humanity's originality, superiority, and autonomy seemed to be in the process of being eroded and decomposed through scientific advance—previously a key symbol of France's civilizing mission. Furthermore, the comparative weakness in France of the feminist (later gay, then multicultural) and consumer rights movements meant that one possible path of resisting "the cold scientific gaze" (while affirming an alternative, improved, and progressive medical practice) and the spread of the market into previously protected domains was simply not considered a viable option. Consequently, and in sum, during this period alternatives to the authority of mandarin paternalism (whether of the state or of professional corporate bodies) remained largely unexplored (and later, openly rejected), in the name of human rights and the human person taken to be universal.

In *Les gardiens du corps: Dix ans de magistère bioéthique,* Dominique Memmi identifies the National Ethics Committee's self-defined mission as one of vigilant surveillance. This vigilance alternately (and in a complementary fashion) took pastoral and martial forms. That which needed protection and defense in France was the dignity of the human person [la dignité de la personne humaine]. It is important to underline that the conception of dignity as inhering in the human person qua existing being is a new one. Immanuel Kant, in his *Foundation of the Fundamental Principles of the Metaphysics of Morals,* articulated the distinctions that shaped the modern philosophic discussion of the term. "In the kingdom of ends everything has either *value* or *dignity*. Whatever has a value can be replaced by something else which is equivalent; whatever, on the other hand, is above all value, and therefore admits of no equivalent, has a dignity." For Kant, dignity was an attribute only of rational beings capable of moral self-legislation, hence of autonomy. Consequently, dignity applied to human beings only insofar as they were rational. "Morality is the condition under which alone a rational being can be an end in himself, since by this alone it is possible that he should be a legislating member in the kingdom of ends."[47] Furthermore, Kant emphatically distinguishes between subjective feelings, for example, *respect* (whose principle is happiness) and objective principles, *dignity* (whose principle is perfection). Finally, for Kant, dignity could not serve as a guide in the search for happiness or for action in any empirical domain.

After World War II, dignity suddenly emerged as the a priori foundational principle of human existence. More research is required to establish how and where dignity had been functioning outside the philosophical and legal mainstream, whose leading thinkers had little to add to Kant for a century and a half. Perhaps the "dignity of labor" might be a place to begin such research. However, until that historical work is undertaken, we can simply underline the dramatic consensus that suddenly emerged. By the end of the war, dignity applied not only to human beings qua rational beings but to human beings per se. On June 26, 1945, the Preamble to the United Nations Charter declared its faith in "the dignity and value of the human person." Strikingly, the same grouping of sentiments and concepts is present as well in the Déclaration Universelle des Droits de l'Homme (Universal Declaration of Human Rights) solemnly promulgated on December 10, 1948. "The human being has a right to absolute protection when it is a question of the respect for the dignity of his person, the dignity inherent to all the members of the human family with their equal and inalienable rights."[48] In this redaction, "human being" is not quite identical to "person," and dignity is metaphorically related to "the human family" and "inalienable rights." Understood against the background of World War II, such phrasings had a clear resonance that commanded immediate assent. An indication that such a consensus existed is found in the fact that the drafting committee of the Déclaration Universelle felt no need for detailed discussions or debate over the meaning of the terms "dignity," "reason," or "freedom."[49] Finally, Pope John XXIII's 1958 encyclical "Peace on Earth" propounded a foundational Christian use of the term.

In France, the task of protecting the dignity of the human person is distinguished by the acuity of its portentous representations of crisis and threat. The National Ethics Committee frequently casts itself in the role of heroic defender of those in danger of imminent, extreme bodily transgression or dispossession. The degree to which justice is portrayed as violated by especially unscrupulous assailants is indicated by the usage of such relatively archaic words as *rapt* [abduction] and *emprise* [seizure], the latter an eighteenth-century term for the violent seizure of land by expropriation. These acts are abuses of sovereign power or marginal brigandage. An additional distinctive characteristic of the French debate and the composition of its ethics establishment is the prominence of psychoanalysts and psychoanalytic discourse. Consequently, it has seemed natural to transfer one locus of debate to the subject and to interpret that subject in psychodynamic terms. Freud, as quoted by committee members, characterized the civilized subject as one capable of distinguishing between *emprise* and *maîtrise* [control], and of making the latter the norm.[50]

The danger of such predatory and/or uninhibited tendencies is exacerbated when the body is at issue—especially a body "at the moment when it is at its most vulnerable, in its most secret part (the gene) or in its most intimate secret (its identity as a being)." Consistently, then, the mandate of bioethics is to "resist possible attacks against vulnerable bodies, whether dependent, fragile, or destitute and handed over [betrayed = livré]" to authorized powers.[51] What could be more vulnerable than the fetus or embryo? One-third of the National Ethics Committee's decisions [avis] concern practices linked to prenatal periods. Another third treat bodies without defenses: sick, dying, those submitted to medical experimentation—"comatose patients without consciousness, demented patients without free will."[52] The committee's maxim might be *Noli me tangere* [Don't touch me].[53] Only two of the committee's texts depart from the theme of aggression and protection: a report and a recommendation advocating the establishment of local ethics committees and the expansion of the scope delegated to the national committee.

Who are the protagonists in this struggle? From whom does the threat come? The answer to the latter question is "the fanatics" [les inconditionnels] who desire knowledge about the body. Three different groups risk falling into this category: patients, scientists, industrialists. As each of these groups' motivations, interests, and positions differ, it follows that the ethical problems they face and the civilizing constraints to be applied differ as well.

Patients need to be protected—from medical and scientific aggression and, to a certain extent, from themselves. Patients' desire for extending the limits of medical experimentation or for distorting national research priorities by placing their needs arbitrarily higher on the national agenda than the needs of others, although comprehensible, must be resisted. As we will soon see, the representatives of a new type of patients group reacted with great vigor and anger against what they saw as the arrogance of the medical and research establishment regarding genetic conditions leading to painful, crippling, and ultimately fatal states.

Scientists and physicians need to be overseen so as to be "domesticated" or "civilized." In France, the ethos of this civilizing mission might be characterized as "physician heal thyself." Although a committee text proclaims: "science has become a concern for society as a whole," almost all of society's guardians were involved in the sectors to be regulated. The missionary and didactic labor required would be performed on one's professional kin. The biomedical elite's call for an accelerated self-civilizing was characterized by a certain benevolent prudence; the desire to surpass acceptable boundaries, in the pursuit of understanding or in attempting to help a patient, were all too familiar to those doing the civilizing. Hence pastoral admonitions were the order of the day.

Lying outside the world of science and scientists, and in strong rhetorical opposition to it, lies a less subtle nemesis, "the market." In commerce, both the norms of the field and the compulsions of the subjects are monovocal: avidity. Here the committee's extension of its mandate to demarcate "the socially legitimate uses of the human body" expressed a large consensus in France. The state must stand between the market and the dignity of the human person. Reaffirming, in the most emphatic of terms, the paradigmatic status of the French position on blood donation, the committee adopted a categorical stance against the commercial appropriation of human parts (cells, blood, genes). Four texts denounce the "parcelization and commercialization" of the human body. The body and person are one: their guardian is the state. "Out of respect for his body, for which he is responsible, but not the owner, mankind must abstain from voluntary mutilation aside from those cases where that act is necessary for the health of the person."[54] Since the individual does not own his body although he is responsible for its care, the integrity of the body—social and individual—is an affair of state.

In sum, the committee's humanism is a complex and contradictory one. As Memmi indicates, "Experienced as a 'forced and artificial' intrusion, since it was decided upon by humans, genetic (or simply medical) investigation can be apprehended paradoxically as more menacing for the body than threats coming from nature (through sickness or malformation)." A humanism that is highly suspicious of human intervention is a humanism that has lost its nerve. It invents and occupies a site that representatives of patients groups and those scientists and health officials with an eye to the international scene would find hard to inhabit.[55]

Sanctification of the Genome

An important colloquium, "Patrimoine Génétique et Droits de l'Humanité" (Genetic Patrimony and Rights of Humanity), was held in Paris in October 1989. The conference was officially under the patronage of the president of the Republic, François Mitterrand. It assembled many distinguished participants, including the leading mandarins of the biomedical and scientific world: Jean Bernard (president of the National Ethics Committee), François Gros (professor at the Collège de France), Hubert Curien (minister of research and technology), and Jean Dausset (Nobel Prize winner and president of the Mouvement Universel de la Responsabilité Scientifique). Its general reporters were Marc Augé (president of the Ecole des Hautes Etudes en Sciences Sociales) and Axel Kahn (a highly placed molecular biologist).

The timing of the colloquium was significant. By October 1989, the American Human Genome Initiative effort was deep into its planning process, with commitments made and scientific and technological strategies under active consideration. In October, James Watson was appointed director of the National Center for Human Genome Research. In May 1988, the French government had agreed to devote eight million francs to genome mapping; a committee chaired by Jean Dausset would decide on its distribution. Earlier in 1989, the United Kingdom officially launched a small genome program, as did Japan; in June the European Commission's Human Genome Analysis Programme was approved in Brussels.[56] Jean Dausset, François Gros, and François Jacob publicly appealed to the Conseil des Ministres de l'Environnement of the European Community in Brussels not to adopt a proposed moratorium on genetic engineering.

The conference's official convener was Gerard Huber, a psychoanalyst and author of *L'Egypte ancienne dans la psychanalyse,* published in *La bibliothèque initiatique.* François Gros and Huber cast the report as a beacon of light in a time of passions, misunderstandings, a priori positions and judgments. What was needed above all else, they asserted, was to establish an ethical foundation that would prepare the ground for directing the coming changes in biology and genetic engineering. "This will not be done without constructing with people (all of us) new social ties demanded by the new situation of biomedicine. These ties must be built on the basis of respect for human dignity and a deepening of the notion of scientific responsibility."[57] Huber presented a synthesis of recommendations of the various committees. These concerned issues of artificial procreation, the limits of human experimentation, brain research, agricultural biotechnology, psychodynamic dimensions of giving life, biotechnological cooperation between countries of the North and South, and philosophic and social implications of these changes.

As for the genome project, Huber presented the following principles:

Ethical Choices:
> Living beings are not assimilable to mechanical machines like clocks.
> There is a qualitative gap [saut] between human beings and all other living beings. This observation justifies the existence of ethical choices but does not dictate, in whatever manner, what those choices should be.
> Philosophically speaking, the ethical speech act should take the following forms: Act and intervene in the genetic text so as to liberate that which, in its "language," makes possible more creativity and liberty in the figurations of speech and action.

Genetic Mapping:
> A map of the human genome must be approached with a great deal of prudence. It must be founded not as a proj-

ect of pure science but as a means of progressing toward
an understanding of a major disease.

Those concerned must have unfettered and free access to
the genetic information.

The project must be carried out without any competitive
spirit. The rules of the market and the economy must
not apply in this domain.

Democracy:

The "democratic deficit" must be overcome; pressure
groups must be regulated not only at the national but at
the European level.

A number of observations are in order. The fanatics remain
unnamed and thereby become phantoms: who is it exactly who
proposed treating the body as a watch? The demarcation of zones
of activity into those characterized by competition, such as industry, and those whose norms are cooperation, such as science, would
stand little sociological scrutiny but does separate out arenas of
good and evil. Finally, making creativity and freedom the defining
human qualities of human beings is not easy to square with the
application of French definitions of "la dignité de la personne humaine" to all human beings, even to "comatose patients without
consciousness and demented ones without free will." Regardless,
for Huber and his colleagues, dangerous, inhumane fanatics are at
large; there are quarantined zones free of their nefarious influence
although threatened by it; the stakes are human distinctiveness.
In this light, it follows that preventive measures must be taken.
In a 1992 presentation, Huber wondered: "Should the genome be
considered the holy of holies, and consequently all access forbid-
den except with precautions and constraints of a spiritual, even a
ritual type, or should it be totally given over to human will?"[58]
Pondering the adoption of spiritual precautions as a defense
against the threat of an unbridled human will might well appear
to be a distinctive recommendation for a commission composed
of the scientific and medical mandarinate of France's Fifth Republic. In fact, it is consistent with the ethnographic context.

Society

It would be consistent with its own self-interpretation to take
French bioethics discourse to be a further step in the "civilizing

process" identified by Norbert Elias during the 1930s.[59] Elias points to Mirabeau and the eighteenth-century physiocrats as the codifiers of a domestication process begun in the court. Its own discursive economy turns on a ratio between civilization and barbarism or decadence. Mirabeau, in his *Ami des hommes,* wrote, "Genuine civilization stands in a cycle between barbarism and a false 'decadent' civilization engendered by a superabundance of money."[60] Two and a half centuries later, the diacritic of a true civilization for a certain elite in France remains the awareness of the need to temper and administer forces of excess while assisting humanity's progress. Its self-imposed task is to create and enforce checks to natural instincts such as aggression as well as what are taken in France to be more social (not sociobiological) failings such as greed and competition. The excessive presence in society of such forces is barbarous, while a superabundance of money remains simultaneously a source and a sign of decadence. "The task of enlightened government," Elias wrote, "is to steer . . . so that society can flourish on a middle course between barbarism and decadence."[61] In the French governmental tradition, the normative project is to constantly make society civilized. Said another way, the substance of French *bios* is never pure *zoē,* mere or bare life, brute or raw drives, untrammeled or primal affects. Or more precisely, this tradition of *bios* has never posited a *zoē* it could not civilize.

What is distinctive about the French situation is both the continuing presence of this normative commitment (albeit taking different forms) and the utter self-assurance that the object of the civilizing process is (always within) the social.[62] Thus, Dominique Memmi closes his superb book by saying that the National Ethics Committee had faced a difficult task, that "of articulating, in a scholarly and secular language, the sacred."[63] For Memmi, the difficulty of the task did not lie in questioning the existence of the sacred, that was a given, but rather in the difficulty of articulating the sacred "in a form that today would be audible to society."[64] That is to say, the challenge was cast as being how to make society speak to itself in a civilized mode about "what is dignified or undignified to do to the human body in the name of great medical and scientific causes."[65] The general point is made directly by Pierre Bourdieu (holder of the chair in sociology at the Collège

de France, France's most prestigious institution), who closes his
Méditations pascaliennes by saying that Durkheim (and French sociology) were not as naïve as some have thought in saying that
"society is God."[66] In France, those authorized to pronounce serious speech acts, simultaneously secular and scholarly, about society, be they sociologists or bioethicists, concur that society is not
dead. Those same authorized spokespeople would insist, however,
that it is under threat. Of course, the champions of the civilizing
process have always held that civilization was under threat. In the
instance under study in the present volume, the threat comes in
a distinctive form; the challenge to *bios* comes from a new instantiation of *zoē*. It is assumed that mere matter must be socialized;
one could say that the mandate of the National Ethics Committee
is precisely to socialize brute life (and the lives of brutes). Lucien
Sève, a member of the National Ethics Committee and an associ-
ate director of the Institut de Recherches Marxistes, approvingly
quotes another committee member, Anne Fagot-Largeault, a professor of philosophy, "The genome is not sacred: what is sacred are
the values that are linked to the idea that we make of humanity."[67]
Whether one finds this claim admirable or irrelevant, it seems fair
to say that "the genome" per se remains exterior to its civilizing
formulation.

The manner in which this specific type of French humanism
poses the problem therefore leaves two major domains of practice
exterior to it. First, the work on exploring, constructing, dis-
covering, and inventing "the genomic" is left to scientists and
physicians (whose striving for advancing knowledge and health
always carries the potential for excess) as well as to venture capitalists and the large multinational pharmaceutical firms. Consequently, the distinct risk is present that the knowledge and truths
about living beings that are emerging in molecular biology and
genomics will be formulated by, and in the interest of, barbaric
and/or decadent forces. Whether this particular humanist dogma,
and the institutions that espouse it, will be able to socialize these
forces afterward remains unclear. There is reason to doubt the
success of such a dogmatic stance for many reasons; perhaps the
most compelling one is that it posits that nothing will emerge from
all this new knowledge that will—or could—radically change our
self-understanding as humans. Dignity will have no further devel-

opments. Second, by definition, the mystical benevolence of this humanistic ethics is not primarily concerned with those to whom its principles are applied. In France, as we have seen, the AFM (and others) strongly contest the view that bioethical principles are more important than the lives of those suffering from genetically based maladies.

This brings us back to the spiritual and to a contemporary purgatorial space. One might say that the spiritual is one product of the way in which current French humanism has answered the question it posed to itself. The appropriate question for an anthropologist of the contemporary to pose about the machinery of the spiritual turns neither on clarifying its conceptual integrity nor on identifying its real referent (we leave these tasks to the philosophers and sociologists). Here, the anthropologist looks for situations in which this machinery is at work and asks: How does such an assemblage operate?

5

Millennium Comes to Paris

Cohen opens the meeting of perhaps twenty people sitting around card tables. The lab heads from the CEPH are present. The Millennium team consists of Chief Executive Officer Mark Levin, several lawyers, and perhaps two scientists.

Cohen says the goal is total clarification of their joint situation. They will begin with an update on the diabetes/obesity work, as that is the centerpiece of the proposed collaboration. Basically he says, "Let's define the issues and see what the answers are." He gives the floor to Philippe Froguel, the head of the CEPH's diabetes project, who is casually dressed in a sweatshirt with an American university logo and consequently in stark sartorial contrast to the coifed and stylish Americans, especially Mark Levin, who is wearing an expensive tie and a silk and light wool suit.

Froguel summarizes the work that has been done at the CEPH, privileging the contributions made by his own team and that of Mark Lathrop. He presents a clinical picture of the diabetes patients under his care. He rehearses what everyone in the room knows, that they are screening for polymorphisms, hunting for the gene or genes involved.[1] From 1993 forward, Lathrop has been looking for evidence of obesity genes using 250 markers. Froguel says it's not too surprising that they have not had much success, as the phenotypical picture is not at all clear. He reviews a number of published studies and indicates his skepticism about their conclusions given the syndrome's fuzzy definition. He is confident that too many different things are being lumped together. Given that messiness, it is impossible to make an intelligent assessment of where they should be looking for candidate genes.

A scientist at Rockefeller University has done the best study so far. He has very good data on 800 individuals. He is doing the first really systematic collection of case histories and hence his data can be reliably interpreted. There may well be overlaps between the obesity studies and the diabetes studies. However, in France, there are significant problems in getting the physicians to cooperate with them. Froguel has made it a priority to train those French physicians willing to cooperate in keeping the type of clinical records that will be useful for genetic purposes. They are making progress and the quality of their collection pool is improving. Nevertheless, they require better methods to characterize the phenotypes. They probably need more markers than they are using. At this stage they wouldn't be able to specify the function or functions of polymorphisms if they found them. Very important questions are: How can polymorphisms be distinguished from mutations? Precisely what is the functional significance of the genetic differences they are finding?

Froguel's team is collaborating on a study of the Druze in Israel. The Druze are assumed to be an ethnically homogeneous population, which should make it easier to identify mutations. However, even in such a controlled situation, with good-quality records, accurate clinical assessment is still needed, and this stage is not going to be automated soon. Hence, data collection is costly and slow. All of this material is well known to those present at the meeting; Froguel is setting the scene for something else.

After this leisurely synthesis of the scientific state of things, Froguel shifts to the proposed agreement with Millennium. Not beating around the proverbial bush, he asserts categorically that any agreement between the CEPH and Millennium must not and cannot be understood as a total collaboration. The CEPH is an academic institution and fully intends to protect its independence. They will remain free to entertain other collaborations. In fact,

Froguel is already involved in discussions with two pharmaceutical giants: Glaxo and Hoffmann La Roche. Pointedly, he says that he wants to make sure everyone understands that each of these agreements will be specific ones; they will not limit other such arrangements. Lest the point be missed, Froguel states that he will make sure he controls his own work.

The CEPH will contribute the DNA it has collected. Millennium will contribute money and its RADE (cDNA technology that displays the expression of multiple genes in different tissues). Froguel insists that there is nothing original about this technology and furthermore it remains unproved. The work will probably take more time than they think it will. (The American CEO is chewing his pen at this point.) Froguel underscores and drives home the point. If anyone in the room thinks that this is a question of Philippe Froguel against the world, they are mistaken. He has been in close discussions with important representatives of the relevant French ministries for some time now—and as recently as last night. He can assure everyone present that the French government will refuse to approve any global agreement between the CEPH and Millennium.

The Americans, not overtly reacting, ask what areas the CEPH is willing to cooperate in. What data will they bring to the collaboration? Froguel responds that there are no proposals they are willing to commit to right away. They have been in negotiations with Glaxo, which is proposing a less restrictive arrangement than Millennium, which stipulates only that the CEPH delay before publishing the data and which does not insist on exclusive control of the data. Glaxo wants to be assured it will have enough time to enable them to jump ahead of their competitors. Froguel has also been negotiating with Sequena, a start-up biotech company in San Diego, as well as an unnamed French company, which has also proposed a nonexclusive arrangement. At this point Cohen turns to Froguel and asks him why a company would want to do that. It is rather astounding that here are the two chief players on the CEPH side getting into a discussion over details that they had obviously not shared previously. Froguel responds that the company wants access to the DNA, that's all. It would be a limited arrangement. Cohen asserts that he is the scientific director of the CEPH and Froguel had better show him written proposals for any such arrangements before signing them.

An American lawyer asks Froguel if he intends to publish the linkage results. Froguel says it depends but eventually "yes." It all depends on the state of the competition. If there is a narrow linkage, then obviously it would be ruinous to publish it quickly, because others could go after the gene and profit from all your work to establish the linkage. What Froguel has to offer is probably a two-year head start on other researchers because of the quality of the DNA collection he has and the linkage studies they allow.

The Americans assert that it is small therapeutic molecules that constitute the real target. The big pharmaceutical sharks are waiting for small molecules. Finding genes is only a step toward small molecules.

Cohen, still mulling over Froguel's early remarks, says he wouldn't trust a nonexclusive agreement.

One of the Americans says that they intend to build a reputation on the quality of their science. It might well be twelve or fifteen years before there is a successful product—that is to say, before they arrive at a therapeutic molecule that was approved by the Food and Drug Administration. This combination of first-rate science and potential profit is in no way incompatible, quite the contrary. He repeats that genes are not their real goal. Genes are valuable, but ultimately they are research tools that can be used to arrive at therapeutic molecules.

The CEPH's Jacqui Beckmann, who had been actively involved in the negotiations, intervenes. He wanted, he said, to raise an ethical concern: how can the CEPH retain control over and credit for the scientific information that they discover? What is the role of the families and the patients organizations who contributed the DNA and made these studies possible? The CEPH will certainly be criticized in France for entering into a business relationship, so there must be a precise and convincing answer to the question of how this will benefit the families. After all, having good family genetic material is the core of this research.

Cohen replies that the academic world is even more proprietary of its research data than the business world. Without proprietary agreements there would be no upstream money and hence no research. Cohen agrees that the CEPH must preserve its identity in whatever agreement there is to be. Cohen insists that France has the least propitious *cultural* environment for the development of genomics and industry. This is a problem.

The Americans switch the topic to some of the advantages of their technological systems. They are excited about the RADE improvements they are making. They are also gaining excellent experience with transgenic mice. They intend to improve the procedures for identifying candidate genes through more use of transgenic mice.

Beckmann insists again that the CEPH has done a lot of work and they cannot simply transfer their material and know-how to Millennium. He underlines that the "candidate-gene" approach based on positional cloning is highly promising and they are working hard at automating it and improving its efficiency.

The Americans agree but make it clear that the CEPH's most valuable possessions are the data on the diabetes families and the genetic material they have collected. This information and material provide a head start of from six months to two years. (Froguel insists it is two years.)

Now the discussion shifts once again to the French side. Froguel says he is against exclusive agreements. Danton, during the French Revolution, was being paid off by everyone and consequently he was independent. Cohen jumps in and says "yes" but remember he was killed by the end of the Revolution—as was Robespierre, who stayed pure.

A discussion then ensues about who will get the credit for any gene discovery that takes place and how that credit will be turned into money and back again into scientific credit. Someone remarks, "We are doing our own psychoanalysis here."

The Americans insist they have no problem whatsoever with Froguel or anyone else receiving full scientific credit for any work or discoveries they may make. This is not an issue for them. Nevertheless, they need to turn that credit into market credibility.

Froguel insists that the CEPH is vulnerable to a range of ethical criticisms. Cohen has many enemies in France and they would be happy to seize any available chance to get him. He has talked openly of the possibility of a preimplantation embryo lab for genetic work, which stands squarely against the formal advice of the National Ethics Committee. He now could be seen as selling the CEPH, of using public money to produce results intended to serve public health and selling them to American businesspeople. Because Cohen is a member of the CEPH and a member of Millennium's board, as well as owning stock in Millennium, he has a conflict of interest. "Conflict of interest does exist! He is going to marry himself" [Il va se marier avec lui-même].

Cohen angrily retorts, "Threats won't work!"

Mark Levin, the CEO of Millennium, intervenes. For this collaboration to work, there must be an exclusive agreement with Millennium about the markers that the new technology identifies. The DNA itself does not have to be exclusive.

Froguel forcefully reminds everyone that the CEPH is not a private research institution. It had become a *fondation d'état,* a tax-exempt foundation of the French state, and hence is officially part of the French patrimony.

More sallies and ripostes and the meeting ends.

Philippe Froguel's Reservations
(February 3, 1994)

Froguel is perfectly willing to collaborate with Millennium or anyone else. He has been involved in the negotiations (traveling to Boston with Beckmann and others on more than one occasion) and is involved in other ongoing negotiations with other companies both big and small, old and new. The point is not the purity of scientific research or nationalism or anything very lofty; the point is this collaboration is a bad deal for the CEPH and—more important—for him. Cohen is handling the whole affair very poorly. Although he is one of the founders of Millennium, he has not included other people from the CEPH (in the scientific or financial planning and rewards) and therefore they do not trust either him or Millennium. Further, Cohen oversells Millennium, as if they had everything in place and their technology would rapidly solve all the problems of identifying disease genes. Millennium's research plan is overly optimistic, especially in their estimations of how many genotypes they can actually produce; they don't know how effective this RADE technology actually is. There is no reason to believe their predictions about candidate genes.

Millennium's first mistake was to start the discussions with the CEPH with their lawyers. This tactic was a bad misreading of the situation; there had been no preliminary scientific meetings and there should have been in order to build some level of scientific credibility and trust between them. Without that trust and that scientific credibility, it was very hard to work with them.

Geoff Duyk, a young Harvard physician and Ph.D. and a member of the Millennium delegation, agreed that the whole collaboration, and even the way Millennium proceeded with its scientific and technological setup, was upside down. They were not yet a scientific entity. Their technologies, while promising, were not yet in place and working. The requisite cDNA libraries were not yet established. There were as yet no DNA family collections. He was utterly confident that all of these things would be put into place quickly but it had not yet been done. He was quite straightforward, "We want a collaboration with the CEPH because we need access to the [genetic] material."

The discussion is all quite amicable on the surface at least, "among scientists" as it were. Froguel replies that if he had to decide about a collaboration with Millennium and Cohen in the near future, "it is no." "We come over to Boston and we see empty labs; you have money but haven't done anything yet. The CEPH is among the world's leading labs and we want more for what we have done. We have never received a scientific argument for the approach Millennium is taking, only engineering plans. Why are all these documents written by the lawyers and not by the scientists? The CEPH has sent scientists over to Millennium for discussions but all the responses have been written by lawyers. We can only suspect that you don't have a real collaboration in mind but only a strategy for robbing us. Send us a scientific proposal and we will read it and discuss it with you. We need to know we are equals in this arrangement. We don't trust Cohen to speak for us. We need academic assurances that we will be able to publish and get credit for our work."

Duyk replies that "we all come from an academic culture and

our premises are the same as yours; as academics our only real resources are our intellectual property. Quite frankly we could be competitors or collaborators. We will all have to decide soon."

There is a profound cultural slippage around the term "collaboration." The Americans keep insisting that it means pooling and sharing. This is not the French view of the matter. They want a detailed statement of who does what, when, and where. They want a plan for how everything will be divided a year from now. Power and delegation of power are crucial dimensions.

Another element of the cultural background that the Americans in good faith were taking for granted is provided in an article titled "National Institutes of Health, Panel Proposes Guidelines for Industry," in the News and Comments section of the February 4, 1994, issue of *Science*. The piece reported the recommendations of a committee whose formation stemmed from a deal causing congressional ire in which Sandoz Pharmaceutical agreed to pay Scripps Research Institute $300 million over ten years for exclusive rights to exploit what was expected to be more than a billion dollars' worth of taxpayer-funded research. The committee drew up tentative guidelines for collaboration between private firms and those whose work had been supported by public funds. After a contentious two-day meeting, the committee put forward some general principles: Participation of an NIH-funded researcher with industry should be voluntary. The right to scientific communication should be protected, and there should be no restriction on future scientific activities. The industrial partner should have a "one-shot, limited-time" option to commercialize research covered by the agreement. It should be on an invention-by-invention basis rather than covering whole areas of research. Corporate partners should be required to exercise "due diligence" in commercializing a product and not lock it up. Because of the multinational status of many of the firms involved, exclusive preferences for U.S. firms should be broadened to include potential jobs generated or other economic considerations.

Millennium met all these criteria. Leading with their lawyers was consistent and coherent for the American context and could be interpreted as a sign of honesty: "everything would be kosher." By 1994, lawyers and businesspeople were full-fledged partners in American bioscience. The situation in France was different;

the academic and medical hierarchies were entrenched and quite comfortable with their power and distinction.

The Research News section of the March 4, 1994, issue of *Science* contained a column titled "University-Industry Collaboration: Huge." The item summarized a paper given at the American Association for the Advancement of Science meetings reporting on results of 1,058 university-industry ventures with research budgets of more than $100,000 that existed on 203 campuses in 1990. Of the $4.29 billion, the researchers estimate that $2.66 billion was spent specifically on R&D—a total that overshadows the previous year's National Science Foundation research budget of $1.69 billion. Moreover, these combined university and private-sector ventures employed 12,000 faculty and 22,300 other doctoral-level research scientists, or 15% of the nation's academic science and engineering workforce. "We do see a weakening of academic norms about the flow of information." Of the high percentage who replied to the survey, 35% said companies can require that specific commercially important information be omitted from a scholarly report; 42% said communication with the public on their research can be restricted. Normality.

February 1994

I go upstairs to chat with Froguel and Beckmann. Their joint office has no windows, only a skylight (more than the computer people next door have), and is a strangely carved-out, almost hexagonal room, cluttered with papers scattered along the tables and desks which crowd the room, two computer terminals in use (a big Unix and a Mac), another sitting on a table but apparently not in use, and a blackboard. This space, like so many others at the CEPH, is heavily used and has been "squatted" by numbers of different people for different purposes. It is heavily trafficked, very little privacy. Even during this most violent of confrontations, the door wasn't closed. This situation is quite typical.

They outline their respective interpretations of what they imagine Cohen's plans for the CEPH to be. Froguel thinks that everything Cohen is doing is part of an elaborate scheme to feed data and resources to Millennium and thereby to profit Daniel

Cohen. At this precise moment, Cohen walks in, open white shirt, mustard-colored jacket, charcoal linen pants. There is nowhere to go. Froguel is wearing clothes he says several times his wife has bought him: a gray linen shirt (with two buttons open and the plastic that the price tag was attached to still poking out), an expensive woven yellow sweater, and jeans. Beckmann looks as usual: jeans, workshirt, sweater, trimmed hair and beard (no mustache), glasses with broken frames, a tired smile. He remains largely silent throughout. Several hours of mutual attack and counterattack ensue, no quarter given. Very little was new in what was said but it was said directly and aggressively face-to-face, with a waxing and waning of tone and energy. Froguel was on the defensive throughout most of the encounter. He kept trying to say very little, both because he was not emotionally prepared and because he knew it was best to keep quiet. This proved not to always be possible for him, and from time to time he would lash out with an insult or particularly nasty piece of sarcasm. His main line of defense was to say time and time again, "I want it in writing." This was mainly in response to Cohen's insistence that Froguel had to leave the CEPH and that the DNA had to be handed over to the CEPH. Cohen said time and time again that he was not ready to put anything in writing: first, because this would close off all possibility of negotiations and would box them all into positions they might not want to hold indefinitely, and second, because there would be letters at the appropriate times but they wouldn't come from him but officially from the CEPH. Back and forth, back and forth. Froguel lashed out several times about Cohen making money, being vulnerable to press attacks, and being a "manager" and not a scientist. Cohen repeatedly asked: Where is the actual DNA? Froguel responded that it was his work. There was one copy at the CEPH and one at another INSERM location. He wouldn't hand it over. He conceded he would eventually but not right away. The existence of the duplicate set was discussed less but its existence loomed over the skirmishes.

Cohen's main line of argument was that Froguel should leave the CEPH with an appearance of amicability, that he would never attack him scientifically, that there was plenty for him to do, it was bad for all concerned to make too much of a scandal of this, etc. Cohen had basically decided that the best way to proceed was to abandon the diabetes/obesity project. Clearly, it was a source of endless conflict; no one at the CEPH could really take it over; the people at Millennium and elsewhere would outdistance them on it. His overall conclusion (drawn over the weekend) was to follow up on Beckmann's idea to put the DNA into the *domaine public* (with conditions). This move would make the CEPH look good to the public and in this way they would get some profit from the DNA, whose value was short term in any case. There was a general movement in the direction of decommercializing primary scientific material. (The NIH had just announced that they were abandoning the patents for cDNA fragments. There has been a change of regime in Washington, but this decision also dovetails with what is apparently a growing consensus in industrial circles that these patents were not worth that much and that even the real gene was only worth 4% of what the investment in a potential therapeutic project would be.) Hence, Cohen was unblinking in drawing the conclusion that there was more to be gained than lost by the CEPH by putting the French DNA into the public domain.

Cohen knew that he had to answer both Froguel's charges of personal gain and the nationalism issue, which the press (and perhaps the government) would focus on. Cohen had scheduled a meeting with representatives of Rhône-Poulenc (the largest French pharmaceutical company) for the next day, to explore possible investment in the CEPH's cancer project. The point of the meeting is to take the wind out of the sails of the nationalist tempest that he suspects is brewing. Cohen repeats one point several times: given that these big pharmaceutical (and biotech) companies are capable of investing vast sums of money quickly ($80 and $40 million), there is absolutely no way that the CEPH—or any public institution in France—can possibly compete with them. For Cohen, only one strategy exists: "join them." The remaining question is tactical: which projects and under what terms.

FEBRUARY 1994

I have dinner at Froguel's house. He is resigned to losing his fight with Cohen but feels it is worth doing. He has been offered a large amount of space for his team at the Hôpital Cochin by Axel Kahn (a highly visible geneticist, member of the National Ethics Committee, frequent intervenor in the press, and reputedly one of Cohen's most ardent competitors and detractors); hence, Froguel has a fallback position. He is nostalgic for the good old days at the CEPH, but that era ended with Millennium. He thinks Millennium is a sham and that they are seeking control of the science. Froguel, as always, is blunt. In addition to mentioning important people by name in the scientific world, he states that he has been talking with a number of government officials, among them Alain Pompidou. Froguel has excellent, high-level political connections. He has worked closely with the Socialist Party (including drafting a report on the "affaire du sang"). Froguel asserts, without irony, that the contaminated-blood affair offers an exemplary instance of what happens when money, science, and politics mix.

During the dinner, his wife (who is of Tunisian-Jewish descent, a graduate of the super-elite Ecole Nationale d'Administration, and the deputy mayor of the left-wing suburb of Bobigny) defends all the accused actors (especially M. Garretta) for their role in the affair. "He was only doing his job." "The American test did not work in any case." The real culprits are the parents of the hemophiliacs, who made it a scandal by going to the media.

Froguel repeats a story (which he has told me several times): that Dausset was somehow a collaborator during World War II—a story categorically denied by absolutely everyone else and for which no evidence whatsoever is provided. This accusation constitutes the purest form of slander conceivable in France. Froguel puts it forth, like so much else he does, in a boyish manner, which makes it hard to take seriously. Such remarks are one thing in private in France and quite another when they become public.

JACQUI BECKMANN'S PROPOSAL

In the heat of the fight, Jacqui Beckmann drew up a "Charter" for the CEPH as a general proposition on how to handle genetic material. In a letter of March 11, 1994, to Dausset and Cohen, Beckmann proposed a draft project as a step in formulating a "general model." First, Beckmann defined what he saw as the CEPH's current mission: the fight against multifactorial conditions and the identification of the "guilty" genes (paving the way to a predictive medicine, to a detailed comprehension of physiological processes, and to a curative medicine). Second, he identified what he saw as the major obstacle: not current techniques but the insufficiency of family DNA banks. To achieve the quality necessary for these banks, one must be rigorous in the nosology and avoid the all-too-common imprecision in clinical definitions; the organization of the bank is a herculean task. The work can be done correctly. One must be very careful now not to let these banks of DNA become privatized. They belong to humanity.

Three principles are applicable: the principle of the sharing of resources between the patients and their families and the clinical and the engagement of the CEPH to ensure access to and the maintenance and improvement of these materials and data.

There would simply be a time delay (maximum six months) before materials are made public. Essentially, the CEPH would be expanding and extending the role it had played in the HLA workshops and then with the "CEPH families." The patenting issues are very delicate: after negotiations between the concerned parties, a government arbitration panel would be created. There would be a fee for access that would vary and ultimately pay for the operation. Ultimately, the scope of the databases would be expanded. Such conditions as well as the identification of the genes responsible for them would accelerate research and understanding of multifactorial genetics.

Beckmann's proposal would make the CEPH the center of world enlightenment in an age of biopower. Several flaws: why would anyone—private or public—contribute their materials? Who would pay for the collection of the data? Patients organizations? Was it plausible to represent the French government as an equitable and efficient arbiter? The CEPH itself played the pilot role in allying with the AFM and creating Généthon, demonstrating a new mix of support, a new model of doing science. During the crisis of the next two years, Beckmann's model will constitute, at least tacitly, the "alternative." Both Cohen and Froguel agreed that it sounded good but was naïve and incompletely thought through.

PATRIMONY

One term that was used sporadically during these events and that might well have been worthy of more extended ethical, political, and legal reflection is "patrimony" [le patrimoine]. Although the French government ultimately did not officially adopt the position that there was such a thing as a "genetic patrimony of the French nation," those who positioned themselves outside this immediate political fray, like Jean-François Mattëi (responsible for drafting legislation), underscored that there was a "legal vacuum" [vide juridique] surrounding the definition and ownership of the biological materials at issue in this affair. As had happened with changes in environmental law in France, there could have been innovation in defining and protecting the status of supraindividual natural materials and information. In the domain of genomics, such innovation did not take place.

Patrimony: what might it mean? According to the art historian André Chastel, the word is an old one, and the notion of a link between valued things and human groups seems timeless. In Rome the term had a precise definition that linked (through time) a legal entity (the family) and the things which belonged to it. This basic meaning of "patrimony" shifted over the course of the centuries as the term's referent was extended on both the subject and object sides. After an erudite overview of those changes, Chastel concludes that the term's modern usage is neither precise nor eternal: "It is a global notion, vague and ever expanding at the same time."[2] This claim is overstated, as a good deal of legal work has been devoted to defining and redefining the term's application in the last two centuries. However, Chastel is certainly correct that its metaphoric and polemic uses are promiscuous.

The modern concept of "patrimony" appeared in the wake of the French Revolution. It was only during and after the revolutionary events that a clear idea emerged in France that before the French Revolution the twin sources of France's patrimony had been the Church and the monarchy. The individual images, objects, sites, and spaces that had been parts of a different whole took on specific value, representative of a highly ambiguous historical relationship (e.g., as signs of a hated regime). Once decontextualized, the goods and symbols of authority were transmogrified into art, culture, history, and "goods." And there, at least once removed from the practices and structures in which they had once been fought against, they became signifiers whose signified was no longer the hated monarchy or the rapacious and obscurantist Church but gradually became shreds and pieces of the nation. In a very short period of time, destroying or selling them became problematic. As early as 1794, "patrimony" was lexically opposed to "vandalism." Barbarians were within the gates. Vigilant civilizers set up committees of defense.

This modern tension between value and culture (and/or value and society) is captured in an oft-cited 1832 quote from Victor Hugo: "Regardless of property rights, the destruction of a historical and monumental building should not be permitted to those ignoble speculators whose interest blinds them to their honor. . . . There are two things in an edifice: its beauty and its utility [*usage*]. Its utility belongs to its owner and its beauty to everyone; consequently no one has a right to destroy it."[3] The invocation of "rights" (and its migration between individual and collective registers) is highly characteristic, as is the opposition of the basely utilitarian and the elevated spheres of value (in this case "beauty"), still linked to aristocratic values like "honor." In conclusion, Chastel ironizes that there has been talk of a cultural patrimony, of an ecological patrimony, and even, of late, of a genetic patrimony. He hypothesizes: "Perhaps this evolution translates the unease of the collective consciousness in the face of menaces, more or less specific, more or less obscure, to its integrity."[4] Chastel's formulation does seem to apply to the case at hand; it is hard not to see that the invocation of "genetic patrimony" fits snugly with the main symbols of French bioethics: menace, integrity, identity. But who reacts? Who is this "everyone"? This "collective consciousness"? The sacred appears again: but is France a primitive society?

An article titled "L'avenir du patrimoine," published in the left-of-center Christian-Socialist review *Esprit,* proposed a reform for patrimony. Such programmatic articles are common in *Esprit;* they provide an agenda for discussion and debate offered to a specific intellectual and political circle in France. That circle includes many members of France's bioethics community. The author, Martine Rèmond-Gouilloud, is a professor of philosophy and an active participant in bioethics debates. She writes from an avowed Christian perspective. Her purpose is less to ironize over the term's fluidity than to contribute to the elaboration of a more self-conscious, and consequently useful, understanding of patrimony. She readily admits, but does not regret, that the term is a vague one. Indeed, a central component of her thesis is that the term's fluidity, its ability to incorporate changing referents, is of its essence. Her thesis turns on this insight; the term's flexibility is precisely why patrimony has a future. Like Chastel, Rèmond-Gouilloud thinks that a renewed interest in patrimony's meaning signals a shared sense that changes are taking place, changes that are experienced as threatening to French identity.

The history of patrimony extends from "stone to life" [de la pierre à la vie], from the stones of the churches and castles of the Old Regime to life itself. Patrimony's trajectory has been a long one, extending from the Roman family and its local gods, described by Foustel de Coulanges, to the French nation and its complex ties to its patrimony, and perhaps, today, beyond the nation to humanity. To write a history of patrimony, Rèmond-Gouilloud argues, is to write the history of dignity. However, to write a history of the forms dignity has taken, to see it as a historical phenomenon shaped by a collective conscience, is to realize the normativity of the term. To fully understand patrimony means to grasp the necessity, from time to time, to reform it. Although the details of French legislative history need not concern us here, that history can be presented as a gradual extension from individual monuments to the physical context within which the monuments are set. Significant among these redefinitions of patrimony, of what needs to be preserved and protected, was the extension of the 1913 law classifying and protecting monuments to the site around the monument in 1930 and then to the monument's immediate setting in the 1943 Vichy law, making it legal to eliminate anything that obscured the visibility of the monument and its site. Finally and consistently, patrimony's purview was extended in 1969 to include dimensions of the surrounding site that were not visible. By so doing, patrimony passed a new boundary; it became future oriented.

In France, the primary arena of application of this new understanding of patrimony has been ecological. Nature is being further integrated into culture.[5] The idea is to keep open a "reservoir of possibilities: postulating a measured utilization of resources, a usage that does not destroy the stock, that ensures its perpetuity." Isn't it the moment, Rèmond-Gouilloud wonders, to extend patrimony further yet? Is it time to extend patrimony to the international level, thereby overcoming "sovereign revendications," and then to all of life? What is life today if not "our genetic baggage, our essence?"[6]

Lest such a formulation sound too materialistic, Rèmond-Gouilloud underscores a fundamental principle: the coherence and dignity of things are not found in the things themselves. Value is human; value unifies what would otherwise be a collection of things. In this sense at least, French bioethical thinkers preserve the Cartesian heritage. Ultimately, a thing is still an object. Although there is a temptation to make a patrimonic thing into a subject (after all, it is valued and cherished), to do so, Rèmond-Gouilloud argues, would be to commit a fundamental category mistake. Things themselves have neither morality nor rights. That is why things need representatives. If patrimonic things are not subjects, they are not simple objects either. Things within a patrimony are both property and not property. A dilemma of bourgeois culture is that as things enter the commercial sphere, they can be exchanged and sold; they lose their distinctiveness. One of the functions of the institution of patrimony is to provide a means of bridging the domains of property [avoir] and being [l'être]. Although value is present in both domains, it is the singularity of the thing's being which "confers its finality and thereby its dignity." The source of that singularity is the human person [la personne humaine]. As the eminent legalist Jean Carbonnier wrote, "It is a constant: property attains its meaning only with the impregnation of the object by the person."[7] Value, however, is assigned by the collectivity. Such a perspective explains why it is perfectly coherent for patrimony to keep changing referents; society is in history, in a perpetual process of change.

Finally, patrimony entails a pastoral attitude, one of care. Since "care" is itself a part of historical process, it has taken different forms in different epochs. Previously the task of patrimony had been dutiful transmission of goods; today it is protection. It is protection in a strong sense because what is at stake, what is to be cared for today, is more than just things, however valuable and valued those things might be. The challenge of patrimony today is to make possible "the reconciliation of man with the world of things."[8] That reconciliation takes place through our exercising care toward things, not by mastering or dominating them: "To take up absolute rights over nature is a beastly attitude." This pastoral turn has produced an unanticipated good; it recovers a lost identity. Contributing to the patrimony is not a disinterested act of altruism. It would be an altruistic act only if the thing contributed to was separate from the individual contributing to it. However, the key point is that those who contribute to a patrimony are not outside it; they are part of it. Once one sees this link, one understands that contributing to the patrimony is neither altruistic nor egoistic; rather, it is "identitaire," a question of identity.[9] Identity is the relational *élan*. Patrimony appears to be an instrument of transmission of goods, but actually it is "the instrument to transmit the urge to transmit. At times one needs to invent reasons for living, for going on. Hence, patrimony."[10] Patrimony is entirely subjective.

This schema raises many questions. Who should exercise sovereignty over the patrimony? Who should be the steward of the world's genomes? What kind of links would this extended patrimony provide? Is it credible that this model will bring Swiss multinationals into concord with French patients groups and American venture capitalists? Are dignity and the secular sacred likely to provide norms that can do what their proponents hope they will? Most important of all, I believe, could such a model capture and represent existing practices in such a manner that one could say a new form will have emerged? The answer to the last three questions has clearly been "no."

"Affair" or Crisis?

Jean Dausset took the lead in attempting to bring the crisis under control. Dausset felt strongly that he had proceeded in good faith, had played by the rules, had entrusted the CEPH's fortunes to the responsible bureaucratic parties. The government (as well as the AFM) had been kept fully apprised of developments and had not evinced any significant warning signs or even indicated displeasure. Dausset knew that Cohen had gone out of his way to put the case personally before appropriate government officials, thus implicating them in the decision, so that later he could not be accused of doing anything behind their backs. For Dausset, the honor of the CEPH was on the line. He had a clear conscience. Albeit with a certain trepidation, Dausset entered the political arena in a forceful, yet dignified, manner.

On March 3, 1994, Dausset sends a letter to François Fillon, the minister of higher education and research, with a copy to Alain Pompiduou, advisor to the prime minister on scientific research, asking for guidance on "the future of the banks of cells and the DNA of the patients and their families." Dausset underscored that he was aware of the novelty of the issue, thereby making it clear that he was not challenging the state's authority but petitioning it, as it were, to take up its responsibilities.

On March 4, a journalist from *Le Canard Enchaîné* comes to the CEPH to interview Cohen. Cohen invites me to stay and record the exchange. The journalist seems surprised but does not object. An energetic and lively exchange ensues in which Cohen forcefully defends his actions. The journalist is low-key, does not ask many questions, and only occasionally counters Cohen's comments with requests for clarification. No fireworks.

On March 7, Beckmann and Rebollo both agree that an underrated aspect of the multifaceted crisis at the CEPH is the departure of Mark Lathrop. Oxford will now be a very important competitor of the CEPH. The new center will concentrate on analysis of genetic data aided by a very strong computing facility. The CEPH is making a mistake by not taking this new situation into account. They concur that Cohen had had a kind of breakdown [démission] during the two-month period when the Millennium crisis peaked. He simply disappeared. No one knows for sure whether Cohen had ever even read the proposed contract between CEPH and Millennium around which everything swirled. If he had read it, then Froguel's interpretation of the situation gained in plausibility, but if he had not, the implications could be taken to be even more damaging. Beckmann drew an analogy with a general or political leader who breaks down during a crisis. Such leaders must take some responsibility for their inaction. Cohen has not done this. Beckmann's tone is irritation more than rage or panic. He is not so much questioning authority as expressing irritation that Cohen was not exercising it.

Froguel, after a good deal of consultation, sends a letter, dated March 7, 1994, on CEPH stationery, to Prime Minister Edouard Balladur. Froguel provides Cohen with a copy of the letter (as well as one for the anthropologist). The letter was drafted to provoke a reaction and it succeeded.

"Please allow me to appeal to you," the letter opened, "to save the French program of the genetic study of diabetes and obesity." The project, Froguel continued, was unique in the world in its scale (5,500 people and 800 families have freely participated as well as 250 physicians and several dozen researchers). The program was exemplary in its exceptional results (several genes identified in two years), which should permit new effective therapies in the coming years.

The project was "menaced" by "the will of the directors of the CEPH to suppress the lab and throw into the street 25 researchers" because of the lab's opposition to an exclusive contract proposed between the CEPH (largely funded by the French state) and an American venture-capital company. This company, among whose founders is Daniel Cohen, has as its goal the identification and patenting, for its exclusive profit, of the genes for diabetes and obesity using the donated material and collected data [les données recueillies] from French diabetics. Thanks to the intervention of the Ministry of Research, the signing of this contract and the transfer of the "banks" of DNA and clinical facts to Millennium and the disbanding of the lab have been blocked.

CEPH's managers decided a few days ago to suppress the laboratory (which produces 50% of the CEPH's scientific publications) and to sell the banks of DNA. "We ask for your intervention to reverse this action, which is against the interests of French medical research but which above all else will ruin the hopes of the diabetics and obese who gave us their confidence."

The reason for suppressing the team is the desire to give away to Millennium, without any control or recompense, "this national patrimony" (the collection of DNA from diabetics and centenarians). The exploitation of this DNA will be financed largely by a multinational corporation. The consequences for French medical and pharmaceutical research will be important and there will be no guarantee of help to the diabetics and the obese: "The essential thing for this type of company is not, contrary to a traditional pharmaceutical laboratory, to develop new medicines, but to acquire a commercial value that will raise the price of its stocks."

We ask for your urgent intervention to protect the national "patrimony" [pour protéger ce "patrimoine" national] at least provisionally. This will allow time for a more considered position to be elaborated.

More than anyone, our team considers that the interest of the patients [des malades] and their families must be a priority over any and all other considerations. We are not talking about "protectionism" because scientific progress has no frontiers.

On March 9, in a conversation with Froguel and Beckmann, Froguel repeats that he had agreed with the principle of collaborating with Millennium. However, over a period of several months, he had repeatedly asked for a detailed plan from Millennium and had never received one. Beckmann, Froguel, and others had spent a great deal of time talking with the CEPH lawyer and preparing a counterdocument and then had simply received no response to it. Cohen had not taken part in these discussions, as he was almost always away. For Froguel, the heart of the matter is the issue of exclusive control of the families' DNA and medical records. Even when Millennium seemed to soften their stance, Hoffmann La Roche (heavily involved in the financing) appeared to be taking a position that they would accept the principle of nonexclusivity but would put so many obstacles in the way of others using the material that it would effectively block competitors from having access to the collections for a sufficient period of time that it would amount to no access.

Who should direct the distribution of the DNA? For Froguel, the response was clear: it should be the patients and their doctors, not the CEPH. In an age of megaprojects, controls need to be implemented to ensure that large-scale collaborations remain public and devoted to the public good. Froguel's discourse tacks back and forth between generalities and specifics. He is doing exactly what he accuses Cohen of doing: expressing abstract humanitarian sentiments to mask his own self-interest. While not exactly agreeing with Beckmann, Froguel has not specified a way to improve on Beckmann's idea for placing the DNA into the public domain for international scientific collaboration, ensuring that individual teams will get the credit for their work, making patients' interests the highest priority, and (here the thought is the weakest) giving sufficient protection to industry so that they will invest the necessary sums of money.

The Affair Unleashed:
Le Canard Enchaîné

The breach into the world of the larger French public came with an article in the insiders' satirical weekly, *Le Canard Enchaîné.* The title literally means "the chained duck"; however, *canard* is slang for "broadsheet." This broadsheet, however, is anything but chained. There is no American equivalent to *Le Canard,* although if one combined the old I. F. Stone insiders' Washington newsletter and the daily gossip from Herb Caen's column in the *San Francisco Chronicle* with *People Magazine* and *The Village Voice,* one might get an indication of its style. It is known as a vehicle in which highly placed insiders leak information that would not pass editorial inspection in more staid newspapers. The Parisian story (which I heard from several sources, giving it the status of "culture") goes that motorcycles wait outside the printing plant, revving their motors, ready to take the first copies to the highest placed officials, bankers, and media stars, who wait in a state of salacious anticipation and fear that their enemies will be featured and not themselves. In a word, *Le Canard Enchaîné* counts.

A headline on an inside page in the March 9, 1994, issue reads "Une Société Américaine voulait se sucrer avec les découvertes françaises sur le diabète" (An American company wanted to sweeten itself with French discoveries on diabetes). From the title to the end of the article, the merely clever skirts the libelous. *Se sucrer* means to get rich but also refers to the problem of sugar in diabetes. The subtitle is "Un patron de la génétique était associé à l'opération, mais les chercheurs et Balladur ont tout bloqué" (A leading figure in genetics was associated with the operation but researchers and Balladur blocked the whole thing). *Opération* refers to a sleazy afffair. Industry, academic research, government—the three estates. Although easy to disinguish rhetorically, relations between these spheres are rather more complicated to disentangle in the world. Cohen could be called a *patron* in the sense of "boss" but not in the sense of the head of a large enterprise. Cohen—and certainly Dausset and the whole heritage of the CEPH—definitely could not be separated from public-welfare-oriented research both "pure" and "applied." Froguel and Pompidou were, in different ways, *patrons;* and politics—and state intervention (real or threatened)—was involved at every level. As we have seen, several of France's most important "ethical spokespersons" had been compromised by nationalism, money, and politics. But the rhetorical device performed a neat division as it was intended to do. One had to be a representative of "industry," "research," or "government."

The short article features a cartoon in which three "Americans," two men wearing glasses and a woman, are watching a television screen. We know they are Americans because one man is grasping a "Coca" can. He says, "Les Frenchies nous ont piqué notre Téléthon . . . on peut bien leur piquer des brevets" [They stole our Telethon . . . we can take a few of their patents]. The two French entertainers on the screen are flanked by numbers indicating the money received, and in the foreground a boy with glasses in a wheelchair beams for the cameras.[11]

The association with the Téléthon rang all the alarm bells it was intended to ring. Massive damage control was hurriedly put into action, coordinated by Cohen and Barataud, the CEPH and the AFM. In fact, the wheels had been turning for some time. On February 25, Cohen had sent a fax to Barataud—with copies to Froguel and the government—emphasizing two points: (1) The CEPH had not "fired" Froguel (it is almost impossible to fire a state employee in France) but was only engaged in amicably discussing his transfer to another institution. (2) Cohen was trying to clarify and protect the status of any and all DNA or cells of *les malades* who had participated in either CEPH or Généthon programs. He wanted to be sure that the materials did not leave before he defined more clearly whatever "specific financial conditions" should apply to them when they were put into the public domain. Cohen was unequivocal that putting the biological materials into the public domain could not be done without the express accord of the CEPH, the AFM, and the clinicians.

For several weeks, the article continued, researchers and Pompidou have been trying to stop the flight of French research results to the United States. The stakes are enormous. Specialists claim that within ten years therapeutic products will be available for many previously uncurable diseases. Hundreds of millions of dollars are at stake as well. In this context, the French team of researchers who found "the diabetes genes" in 1992 have temporarily halted the transfer of their work to the United States. (It is worth underlining that Froguel's coinvestigator, Gilberto Velho, is Brazilian, and the lab employs Tunisians.) What exactly was meant by "diabetes genes" is unclear. Linkage studies from the CEPH and elsewhere had indicated that it was probable that there were genes involved in various diabetic conditions; Froguel was hot on the trail of "candidate genes." But no one in 1994 had identified "the genes for diabetes." The hype about hundreds of millions of dollars and a host of genetic cures on the horizon is exactly the rhetoric of the world of venture capital and sensational journalism toward which *Le Canard* was attempting to whip up disgust.

Starting in 1990, the article continued, a young doctor, Philippe Froguel, initiated a program to identify the diabetes genes with the help of 5,000 charitable workers and 2,500 physicians. This trope of a collaborative effort in the name of public health and under the banner of solidarity is extremely important. Two labs were at work: Généthon and CEPH, the latter directed by the "publicity hound" [très médiatique] Professor Daniel Cohen. In 1992, in Froguel's lab, "three diabetes genes are discovered using the DNA of the donor families." Hence it will soon be possible to have a test for diabetes and eventually a genetic therapy. But it will be expensive. "Une société américaine d'investissement, Millennium" (the terms "start-up" and "venture-capital start-up" had not yet entered into French), proposed a scientific and financial cooperation. The Froguel team was ready to trust them, especially given that one of the five founders was none other than

Daniel Cohen. But things got hairy [se corsent] when the Americans proposed to buy the stock of French DNA and whatever they could discover from it. According to Millennium's proposal, the French researchers would only be able to continue playing with their test tubes if they agreed to send their results to Boston and not to sign contracts with anyone else. Obviously Froguel refused to sign (Evidemment, l'équipe Froguel n'accepte pas de céder ses découvertes ni son DNA). Millennium is a start-up that hopes to make money. Cohen has 2% of the shares. But Froguel discovered the genes, not Cohen—where is his money?

At the center of the dispute is Cohen's assertion that no French company is in a position to do the work Millennium proposes to do. The spokesman for the French government, Alain Pompidou, disagrees: "there are excellent French companies for genetic therapy." Cohen swears the opposite. Pompidou explained that "the state wanted to prevent an agreement that was neither reciprocal nor included an exchange of scientific information. We must also ensure that intellectual property is protected." *Le Canard* sarcastically adds, "Yet another fanatic of cultural exceptionalism." This statement refers to France's attempt to impose the French language exclusively in the media and in scientific meetings so as to stem the tide of American imperialism. Cohen is now proposing to make Froguel's discoveries the property of the international community. The idea is seductive. But Cohen's enemies are not convinced. American labs have patented discoveries made by French researchers. Pompidou adds, "If there are to be relations with foreigners then they should be under clear conditions." Froguel asks the Americans to practice the same *transparence* as the French and to forbid patents to be taken out on genetic material so as not to allow industry to make products from them.

Perhaps the most damaging statement (at least internally within the world of French genetics) is *Le Canard*'s quote of Cohen saying, "L'ADN n'appartient pas aux chercheurs, il appartient aux financiers" [The DNA does not belong to the researchers, it belongs to the financiers]. The journalist had simply cut off the end of Cohen's sentence, which continued "financiers d'AFM." The financiers of the AFM, as the journalist knew, were the patients groups.

Response

On March 9, 1994, members of the CEPH sent a petition to Dausset asking for Froguel's ouster. For the last several months, the petition read, "by adopting an outrageous attitude and a politics of leaks to the media rather than one of conciliation within the CEPH, Froguel has willfully harmed the integrity and image of the Foundation to the outside world. This reality has created a detestable climate within the institution which affects everyone who works at the CEPH." Since Froguel has detached himself [voire désolidarisé], the petition continued, Dausset should make this detachment official. The key phrase is *voire désolidarisé,* which carries the simple but effectively charged sense of a betrayal and also of an attack on "solidarity," a key French symbol of modern French Republicanism.[12] It is important to underline that however fractious the normal life of the CEPH had been, and it was quite fractious, there was essentially no internal division when it came to Froguel's conduct. Outside the inner circle of Froguel's lab, there was basically no support for him. During the coming months, the CEPH spoke with one voice on this issue. The partial exception was Jacqui Beckmann, who was solicited by Cohen to play the role of personal mediator, or *casque bleu.* While partially defending or explaining Froguel, Beckmann never adopted a stance that could be called anti-Cohen. Such a position was untenable in the climate of moral warfare and Beckmann would pay dearly for holding it. Over the next month he essentially lost all authority within the institution.

On March 10, Dausset sent an official letter to Froguel. The letter opened by saying: "The attitude that you have adopted during the last several months as regards both the scientific direction of the CEPH and its researchers has created an atmosphere that is extremely unfavorable for research. It tarnishes the image of the CEPH and has attained such a degree that it is no longer acceptable." It continued, "Your deliberate willfulness in taking individual initiatives without consultation, the systematic denigration of CEPH's scientific directors, the circulation of inaccurate information have led to a loss of confidence that makes it impossible to maintain your activities at the CEPH." Dausset asked Froguel to cease his activities. Froguel was asked not to remove materials from the CEPH. However, access to the DNA of the diabetic and obese families remained "possible under conditions similar to those that could be given to the general scientific community." Given the circumstances—the legal, ethical, and political ambiguity surrounding exactly what normal access was supposed to be— the last phrase is telling. It basically left the door open for the diabetes research to continue; consequently, the CEPH protected itself against possible accusations that it was obstructing progress for *les malades.* Froguel's exact status at the CEPH also remained vague. This ambiguity neither bothered nor hindered him. He continued to come to the CEPH, daring Dausset to publicly bar him.

On March 11, in a physicians newsletter, *Impact-Médecin Hebdo,* in an article titled "Votre ADN m'intéresse" (Your DNA interests me), Froguel posed the rhetorical question "Is there a place for nationalism in scientific research?" He warned that French research was faced with a new GATT treaty dominated by the Americans. There was a real danger of American hegemony over medical research. Cohen responded to the journal's question "How should France's world-class genetics research be protected?" by taking the high ground, "My first concern is for those suffering from these diseases, to help them as quickly as possible."

The article reported that the government had "intercepted a fax" from the CEPH to Millennium and promptly intervened. This espionage trope became a recurrent theme both within the CEPH and without. People were firmly convinced that "the ministries" were tapping their phones, monitoring their e-mail, generally spying on them. More prudently, Professor Jean-François Mattëi (a center-right deputy from the South of France and very active in the ethics commissions) confirmed that there was a juridical void in French law.[13] He reassured the physicians that the government had a commission working on the matter and France should have a policy within two months. Mattëi reassured the audience, "An amendment must be rapidly put before the National Assembly in order to regularize these practices." Such an amendment would have to be cast on the European scale.

In the March 11 issue of *Quotidien de médecin,* Pompidou framed the CEPH crisis as an "economic and an ethical problem" insofar as there were rules of correct conduct between doctors and patients that were in question.

On March 14, Dausset sent Pompidou a letter informing (reminding) him that the CEPH's administrative council had met in October (on the twelfth), and consequently, the government had been fully aware of the negotiations with Millennium. Nothing had been finalized; the agreement called for a bilateral scientific council; the need to rectify financial and other matters was duly noted; ethical questions had been taken into account and it was stipulated that any contract would have to conform to the positions of the French bioethics laws; Daniel Cohen's position in Millennium was clearly announced; the Ministry of Research had been informed and did not manifest any opposition. The minutes of the meeting had been unanimously approved at the administrative council meeting of December 21, 1993. As the council included members of the AFM and the ministry, all parties concerned were perfectly *au courant* of the precise standing of things. Of course, the government knew exactly what had transpired at the meeting. Pompidou's statements to the press are vague and general. For the time being, the press was not highlighting the contradictions between innuendo and specific charges of wrongdoing.

On March 15, 1994, *Libération,* the formerly maverick leftist daily, emblazoned its Société section with the headline "Les dollars jettent le trouble dans la génétique" (Dollars cause trouble in the genetics world). The article, written by their knowledgeable science writer Corinne Bensimon, herself of Tunisian Jewish origin, opened by posing two core questions: Who should profit from the discovery of genes involved in disease? Under what conditions should research funded by public funds be ceded to private labs?

In the article, Froguel is asked to respond to Cohen's reversal over putting the DNA in the public realm. He replies, "How can it be accepted that anyone can exploit this material—including commercial entities and foreign ones at that? Such a move would hurt French research and eventually even the French pharmaceutical industry." Cohen responds: "I refuse to consider health products like electronic products in a commercial war. The interest of the patients is more important to me than the national pharmaceutical industry." The article then quotes Bernard Lepecq, a biotech-nology specialist at Rhône-Poulenc, as to why there should be collaboration between French labs and American start-up compa-nies. The reason is simple: "It is almost impossible to begin a start-up company in France because the capital market is too narrow to find sufficient money for high-risk ventures." The statement refutes Pompidou's claim that French companies were capable of doing this work. Rhône-Poulenc had been the main French phar-maceutical company to enter into the biotech start-up market in France. Company representatives, as we have seen, had recently been at the CEPH.

Although the tone of the article and its headline were inter-preted at the CEPH to be anti-Cohen, the article's content was balanced. The article reported Cohen's offer to make the DNA public, as well as his humanitarian reasons. Most revealingly, the statement from a well-placed official in the state-assisted and very powerful Rhône-Poulenc directly contradicting Pompidou's assertions confirmed Cohen's basic points. Rhône-Poulenc had re-cently expanded internationally, supporting ventures in the United States, and they knew very well that there really were no com-petitive existing French players in this domain. Froguel knew this as well, as he was negotiating actively with both American and British private companies. French stock market rules were such that a company had to have made a profit to be listed on the stock exchange; because all biotech start-ups lose money, this stricture effectively ruled out start-up companies funded by venture capital. In France there had been a few biotech companies formed under the wing of the large pharmaceutical companies but they had not prospered. Finally, Rhône-Poulenc had been in touch with the CEPH about the cancer project, and the cosmetic firm L'Oréal was partially financing the "aging" project. Neither company had expressed interest in the diabetes project.

I interviewed Bensimon on March 21, 1994. She began the exchange by asking me if I was there as a spokesman for Cohen. This remark was a curious one. She knew that my friendship with certain journalists at *Libération* predated my dealings with Cohen. She had known Cohen for some time herself. Although it was plausible that he would attempt to enlist my services, it was a sign of a nervous journalist to think I would accept. Her sources should have told her that I had been seeing Froguel as much as, if not more than, Cohen. Finally, as French stereotypes go, it would have been implausible to assume that an American academic from Berkeley would be an enthusiast for American business. I explained my role as neutral participant-observer, adding that from within the CEPH, the situation was viewed as essentially "un combat de coqs," a cockfight. Froguel was trying to dethrone Cohen and was receiving encouragement from sectors of the Parisian scientific world. Cohen was putting down a rebellion and asserting his authority. Dausset was concerned about the prestige of the CEPH. She laughed, agreed, but added that interpreting things in terms of gender was a typical American view that missed the real issues. She thought the loose organization of the CEPH made it vulnerable. Barataud clearly understood that the AFM had to be purer than pure. The CEPH should do the same. The anthropologist is given yet another message to transmit.

She had received a number of anonymous calls from researchers denouncing Cohen for a wide range of nefarious acts. The tradition of *délation,* of informers, is an old one in France (from the *lettres de cachet* under the Old Regime to the tradition of concierge as police informant to the massive wave of anonymous informers under the Vichy regime).[14] Bensimon was not surprised by these acts themselves as much as by the virulence of the opinions expressed against Cohen. She asked me whether I knew why well-placed and secure researchers like Axel Kahn seemed to be so strongly against Cohen. As if to answer her own question, she immediately agreed that expressions like those of Pompidou about "French DNA" were horrible. However, she was reluctant to take a public stance against this base nationalism, a position her newspaper generally opposed when it came to issues like immigration. She agreed that the nationalistic mood being whipped up by the far right was a dangerous one to be associated with but she then changed the subject.

In a press release on March 16, 1994, Dausset, Cohen, and the CEPH's Scientific Council announced their intention of clarifying "the mission that society has conferred upon us. It is a question of putting all our energy, imagination, and creativity into the effort to eradicate the suffering linked to certain hereditary illnesses." Two different issues are at stake: (1) that of the relations of academic and industrial science, including the issue of conflict of interests; (2) the newer issue of the ownership of biological materials given as gifts [des dons biologiques], concerning which there was a legal void. The CEPH has always been loyal to the interests of France, but the interests of *les malades* must have priority. The specifically French ambiguity of the situation is conveyed in the opening phrase—who, precisely, is this "society" that has given the CEPH its mission? Is it humanity? The French state? *Les malades?*

The scandal hits the international scene. The March 17, 1994, issue of *Nature* ran an article titled "French Geneticists Split over Terms of Commercial Use of DNA Bank." *Nature*'s reporter cast his net wider than the French publications. He quotes Mark Levin of Millennium as confirming that a proposed deal with Hoffmann La Roche for work on diabetes and obesity did not depend on the DNA bank. In fact, Hoffmann La Roche subsequently did put $80 million into the project after the CEPH and Froguel had withdrawn. Levin observed that "the wariness of French researchers about commercial ties is reminiscent of the early days of biotechnology in the U.S." He pointed out that, as there is no open access to CEPH's bank, CEPH and the French government were already effectively exercising "an exclusive licence on it." *Nature* quoted other U.S. researchers: "It is not like a patent issue," one leading genome researcher says. "Millennium would get a head start; it would not give them a blocking position."[15] He added that companies need DNA from different ethnic groups and if they privatized or cut off access to these data banks, they would lose them all. Levin's point that what was at stake was not the ownership of biological materials per se but a head start in what would by necessity be a competitive/collaborative relationship was entirely neglected in the official public French reactions. Cohen and Froguel and Beckmann fully agreed. Timing was of the essence, not ownership of the material. Froguel proceeded exactly in line with that logic in private while fanning the nationalist and anticapitalist rhetoric in public.

An article titled "French Gene Mappers at Crossroads" appeared in the March 18, 1994, issue of *Science*. A full-page insert, titled "Dispute over Company Link Roils CEPH," featured a picture of Philippe Froguel.

> CEPH is not alone in confronting these issues. A handful of new gene-hunting companies are trying to forge links with academic labs that hold collections of disease-afflicted families. Like Millennium, several of these companies are asking for a period of exclusive use of the DNA in return for a financial commitment to the lab. "Some grace period will probably be required," says David Galas, formerly head of the U.S. Department of Energy genome project, and now vice-president for research at Darwin Molecular Technologies Inc., a Seattle-based company that is going after the genes underlying auto-immune disease. "Some geneticists argue that this would be nothing really new. Academics, they note, rarely share DNA until they publish a result indicating the probable location of a disease gene. But other researchers—particularly those working on multi-gene diseases, where pooling data from family collections made by different groups may be necessary—are deeply troubled by this idea."[16]

Froguel told *Science* that Cohen "doesn't know if he is working for Millennium or CEPH." Cohen vigorously rejected the charge; he took himself out of negotiations and hadn't read the contract until several weeks ago. The decision must be made by the Dausset Foundation governing board, which includes government ministers. Cohen, the article reported, is now proposing that the CEPH make all of its DNA samples publicly available. "Academics could use the samples at cost price; companies, he suggests, could pay royalties back to CEPH, if they develop a product from research using the samples. The basic idea is not new: Similar public repositories, each containing samples from more than 100 families affected by insulin-dependent diabetes, are being operated by the British Diabetic Association and the Human Biological Data Interchange, which distributes DNA from the Coriell Institute for Medical Research in Camden, New Jersey." Froguel agreed to Cohen's proposal with two caveats: that any group using the DNA must make public its own DNA and data and it must refrain from patenting any results obtained using the CEPH samples. "That would effectively bar Millennium or any other company from making use of CEPH's DNA." Mark Levin told *Science,* "It's our policy to wait for them to decide what's best for themselves."[17]

The March 24, 1994, edition of the prestigious *Le Monde* ran an article titled "Les pouvoirs publics veulent une transparence des recherches sur le genome humain" (Public powers want a transparency about human genome research). *Le Monde*'s medical reporter, Franck Nouchi, posed three questions: Who owns the DNA from patients groups? What is the exact role of start-up companies? What should the role be of such companies with respect to French researchers? Staying on the high road above the fray, *Le Monde* quoted the March 7 letter of Froguel to Prime Minister Balladur, which asked him to "save the French genetic program on diabetes and obesity." *Le Monde*'s gloss is highly significant: "Although legitimate issues are raised, the unease does not entirely correspond to the facts of the situation, at least as they are reported in the minutes of the October 12, 1993, meeting of the Jean Dausset Foundation." *Le Monde* was sending an authoritative signal that there might be a crisis at the CEPH but that there was no "affair." The article marked the end of the public polemics on the CEPH and Millennium.

Le Canard Enchaîné returned to the fray one last time on March 30, 1994, with an article titled "Une première scientifique française: le gène gratuit qui fabrique des dollars" (A French scientific first: the free gene that produces dollars). The subtitle read "Des découvertes fantastiques mais, pour 'le bien de l'humanité,' aucun brevet déposé. Des labos US sautent sur ce filon" (Fantastic discoveries but "for the good of humanity," no patent has been taken. U.S. labs jump on the goody). The cartoon is of a smiling President Mitterrand receiving a blood transfusion on which is marked DNA of "hundred-year-olds," and Mitterrand is saying, "If it works, I will run again." The article opens by quoting Bernard Barataud lamenting that the AFM had not patented any of the genome. He has watched the "commercial fever and financial negotiations" taking place over "the human genetic patrimony." *Le Canard* explained that "the human genome is a sort of 'living book' in which everything is defined: eye color, foot length, nose size, as well as eventual predisposition to diabetes, to dystrophies, to all kinds of diseases, of advantages, and perhaps the secret of longevity." Finding genes is hard to do. But a brilliant French team has done it for diabetes. And an American lab wants to try the same thing for the longevity gene, "if the kind French researchers will only pass them for nothing the collection of DNA taken from our frisky old people."

The whole situation is dangerous explains an unnamed "grand prof parisien de génétique." Others fear not only exploitation but eugenics. Why not "putter with defects" [bricoler les tares] or institute a selection before birth? Given how much money there is to be made, *Le Canard* fears the worst. France has it right: there should be no commerce of the human body. Patents in genetics should remain open, dedicated "uniquely to advance humanity's happiness. In this manner, French research graciously puts its biological materials and discoveries at the service of the international scientific community." The Americans, on the other hand, are commercializing and/or stealing everything. For example, there is a small laboratory in Alabama that sells DNA molecules at high prices, including molecules discovered by French labs. Such procedures have its fans in France as well. Daniel Cohen at the CEPH had bought himself an advertising agency in the United States at French expense.

The article is revealing. First, Froguel and Millennium are basically absent from the story. Second, the examples given are specious: selling reagents and the dispositions of "biological gifts" are completely different matters. Third, the collection of DNA from hundred-year-olds was being financed by a French firm. The article is a parting shot.

The final, and significant, journalistic entry came from the conservative *Le Figaro* on April 4, 1994: "Les banques très convoitées des généticiens" (The highly desirable genetic banks). The article, written by a physician, coolly outlined the story, concluding with the observation that French investors are skittish about such ventures and must bear part of the blame.

6

Normalization

A call from Cohen, touching base. He has not been brooding but acting. He is certainly the *anti–homme de ressentiment*—active, active, active. He feels that the worst is over in the press. Highly placed insiders are soliciting *Le Canard* to lay off. The article in *Le Monde* was not a personality piece but a presentation of the issues. Apparently, *Le Monde*'s reporter has found statements from Philippe Lazard, head of INSERM, and Axel Kahn about the need to work with the Americans dating from before the affair. These documents will certainly make it harder for others to continue taking purist stands on science and nationalism.

A thirty-member government commission has been set up. It has no representatives from the human sciences. There is one representative from industry. Cohen and Dausset now agree that the CEPH should take the initiative and present something to the commission as a way of orienting the debate. This should be a document which becomes a talking piece.

Cohen met with some industrialists who agreed that French capital was out of step and dangerously behind in the biotech arenas. They are going to meet with him soon.

March 25, 1994

The government informed Dausset that a commission had been set up to study the question. The group hoped to deliver an initial report by June 10. Cohen was relieved because it meant the government was buying time, clearly not intending to take any decisive action.

In an editorial in the internally distributed *CEPH-Info* for March 1994, Dausset observed that the "turbulence" at the CEPH at least put some important questions on the table. The "most fundamental" is to whom does the biological material collected by the CEPH belong? "The obvious response is those who gave their blood." However, it would be good to know whether blood was given for a specific problem or general scientific research. If the latter, then "all statistical and anonymous studies would be legitimate." Does the center that collects the material have the right to dispose of the material as it sees fit? The CEPH, in its universalist vision, has always taken the stance that matter given to science should be distributed under appropriate controlled conditions. Although the principles are clear, their application is not. If there is money to be made, shouldn't those who gave their blood benefit? Their nation? Should the scientists who made up the bank of materials be excluded from profit? The institutions at which the work was done? "In this mess, it is above all the interest of the patients that must guide us, not this or that patient individually, but the ensemble of present and future patients. Now it is clear that their interest lies in the most rapid discovery of a treatment or at least a way of tracking that could lead to a preventive treatment. It is up to society to resolve this difficult question on the basis of ethical, juridical, and economic reflection."

With these questions and these noble sentiments, the problem is thrown back to *la société*. Who could be against *les malades*? Who was putting the safety of *les malades* in question? Was it the rapacious Daniel Cohen and his allies *outre-Atlantique* with their dollars? Or was it, as Cohen and Dausset intimated, those who were preventing research from taking place? The villains in the latter narrative would be much closer to home.

MARCH 26, 1994

Spent much of the day with Beckmann. The discussions turned around his scientific work, about which he is concise, crisply making the essential points, as in a lecture. He leads me through the basics of his gene hunt. The most interesting discovery in this "monogenic" malady (one of the muscular dystrophies) is that the "single gene" is found in a number of different places in the genome and that it is not always the same gene that is defective. Hence, although 3,000–5,000 maladies may each be caused by a single gene, as the press and the spokespeople for the genome project repeat over and over again, these genes may be found at many more than 3,000–5,000 locations.

Beckmann is adamant that the CEPH needs to be more democratic. Cohen operates with a "tactical soft tyranny." He separates people; everything operates through him; he is unpredictable; there are no effective intermediaries to curb his whims. Cohen has no managerial philosophy. He operates in a crisis mode that minimizes any institutional constraints on his action. He is adept at forcing or displacing issues. He constantly uses Dausset as the moral guarantee of everything he does. Although Dausset usually does not know all the details, he is complicit with Cohen and rarely, if ever, reigns him in. This system has been a productive one.

Beckmann has a list of grievances. His stance is principled but not political, as he is not willing to organize against Cohen (and often operates on his own seeking special access). Beckmann speaks his mind in public, and like Froguel he is a *cavalier seul*. Beckmann is increasingly marginal at the CEPH because he has dared to say that Froguel is one of the victims of the Millennium events. This stance was taken very badly by several outspoken players who situate themselves as external to the fight and see it as hurting the CEPH. These people demonstrate no solidarity with one another beyond this issue.

Although it had been agreed that there should be a meeting of lab heads on the last Friday of each month, these meetings have not taken place. When it is finally announced that there is to be a meeting of a newly rejuvenated committee of lab heads, Cohen asks Beckmann *not* to be a member of it because it would create too much dissension. "Don't worry," Cohen told Beckmann. "You will be my special counselor."

Beckmann, marginalized for his (partial and critical) solidarity with Froguel, is now systematically ignored by Cohen. Beckmann could well leave the CEPH and work full time at Généthon, where his genetic research is based; he could compromise his scruples and remain quietly at the CEPH; or he could attempt to organize something at the CEPH. He hesitates between each of the alternatives, articulating their positive and negative points but not taking any definitive action. Christian Rebollo says that he himself has *freely assumed* his subordination to Cohen and that Beckmann should do the same.

Froguel, in a good humor, was in the office for much of this discussion. When I asked him if his plans for Oxford were a kind of reply to Millennium, he concurred.

MARCH 29, 1994

Lunch with Froguel. He asked me to outline my analysis of the situation. He agreed with my basic points: this was essentially a personal fight for power and prestige between two strong men; he would have signed on with Millennium if they had proposed an attractive contract to him (Millennium had offered him an insulting consultant's fee of one thousand dollars a month and listed him quite low down in the consultant list at the CEPH, equating him with the technicians; he blames Cohen for this humiliation); the nationalistic dimension is unsavory and dangerous for science in France; there will have to be systematic cooperation with both the biotech and the pharmaceutical industries (i.e., basically Cohen's analysis of the French situation is correct). Froguel insisted that he never formally negotiated with Millennium. They insisted on a strict division between the scientific, the economic, and the legal dimensions. Whenever issues were raised by CEPH scientists, they were told that the scientists at Millennium were not authorized to discuss them. When, during the fall, he put his objections in writing, his letters were never answered.

Froguel knows that one or more of the major players in the Parisian molecular biology world who had promised him support if he took on Cohen might well renege on their promises. There were already indications of this, as the laboratory space he had been promised had not yet been made available. The situation has reached such a point of stalemate that Cohen is now making phone calls around Paris seeking to consolidate the offers to Froguel as a means of having him leave the CEPH as rapidly as possible but in a fashion that will not leave Cohen open to the charge that he sabotaged the diabetes research. At both the governmental and the university levels, Cohen is receiving vague replies: one important personage confided to Cohen without irony that "Froguel is not trustworthy after all."

Who was the big winner in the affair? Mark Lathrop. Froguel and Mark had proposed a kind of European triangulation with the CEPH, Oxford, and a third partner. Cohen and Dausset rejected it in favor of going American with Millennium.

March 30, 1994

Yesterday was democracy day at the CEPH. Cohen called a meeting to "discuss" the plan for the future of the CEPH that he will present to the board of directors tomorrow. That meeting is important because it will both vote the budget for 1994 and provide a mandate for Froguel's ouster from the CEPH. It is vital that Cohen present the CEPH house as being in order. Hence he has constituted a scientific committee whose members' names he cannot remember (he refers to them as "nos amis") as well as this internal "representative" body. He invited certain people to come to the meeting tomorrow to embody the democracy and to testify that Froguel has been a major social impediment to the smooth functioning of the CEPH. These tactics are presented straightforwardly. No one posed major objections to the principles or the overall strategy, although a few minor suggestions were put forward on tactics. Having given them at least the semblance of formal responsibility, Cohen then asked the council to exercise it. They are to confront Froguel and tell him to keep a low profile at the CEPH. This puts them in a police role, which some relish.

The second part of the meeting was to outline the "new CEPH." Cohen's formula is technology + major scourges:

1. The physical mapping of the genome: 50% of the CEPH. The team of Ilya Chumakov, the computer people, the remaining technicians, and perhaps some of the YAC people. There is nothing new here; it is a reaffirmation of the centrality of mapping for the CEPH—to a large extent because there is nothing else to put in its place. There was no concrete discussion of what new technologies would be explored.

2. Service: 25%. The CEPH families; the YAC service (motivation is low); the prostate project, which is just under way; and some of the group working on aging (this is essential in order to gain new space in the Hôpital St. Louis).

3. Scourges: 25%. This means the Belgians' cancer project. Cohen said that the CEPH has always gone after the big issues and that diabetes is not big enough and the payoffs will be too far down the line. Even if Froguel does succeed in finding a new working arrangement, it is unlikely that he will be able to compete with Millennium and others. French researchers in any case will not be dominant in this field, because there is too much foreign interest about the same genes and their commercialization. AIDS is a minor experiment with no immediate payoffs in sight but is not costing the CEPH anything.

Cancer is the "sponge thrown in the water" (a phrase used several times). The Belgians are somewhat mockingly referred to as absolutely dedicated and totally involved in their project, thinking about nothing else. There continues to be a fair amount of by now relatively gentle joking about them. Cohen continues to characterize their project as "too complicated to explain."

This leaves the project on aging, about which very little was said except for Cohen asserting that Schachter was on the path of "Froguelization," thinking his project was uniquely his own responsibility.

Beckmann's project can continue until it is finished, but his is a Généthon project. Because it concerns a monogenetic condition, it can't be the future. Cohen, explaining why Beckmann had been excluded from the leadership group, said, "He is comfortable when he is excluded." The other sharks were eager to cut him out. Beckmann is now "sur la touche," on the mat, close to being forced out of the CEPH.

All in all, nothing very new: the hole created by the loss of

Millennium has not really been filled in, and a brave face is being put on things. Cancer receives a nod, but only three people are involved for the moment. This patchwork arrangement will do for the meeting but Cohen is hardly satisfied. The mood of disenchantment, *méchanceté*, bitterness, and uneasiness over the present and the future that characterizes the general French situation applies in spades at the CEPH. People talk of leaving but for many there is nowhere to go.

April 1, 1994

Apparently the meeting of the board of directors, which was supposed to be smooth sailing, was a rather stormy affair. Although there had been a solemn vow of confidentiality, details were leaked immediately. The government had chosen its strategy: reassert control and exercise its bureaucratic authority through minor penalities. An intense attack was made on the CEPH budget, with a number of items under close scrutiny, sending a clear warning that the research directions of the CEPH would be monitored. The Froguel affair was brought up, and the two CEPH representatives spoke out against him. The CEPH was officially asked to break off contacts with Millennium. Later Cohen said there was a rumor to the effect that the French secret services had intercepted faxes from Millennium to the CEPH. Rumors fly that e-mail is being monitored.

Froguel seemed quiet and chastened today. It was announced that Hoffmann La Roche had invested $80 million in Millennium.

I walk downstairs and Cohen is there. He is morose about the meeting with the board, sparse on details but clear that he considers his hands increasingly tied and he doesn't like it. He tells me about a letter sent to him from French diabetes specialists; it was full of sentences about protecting the nation. He calls one of the Jewish doctors whose name is on the letter, and while "*tu-toyer*-ing" him (and getting his name wrong), Cohen asks whether he does not find the document to be "Vichyist." The letter has red, white, and blue stripes on the upper corner. His respondent becomes quite defensive and gives Cohen the phone number of the doctor who organized the petition.

Later Rebollo says that there are three modes of interaction with Cohen. First, there is a form of vassalage, in which there is also always contempt on Cohen's part; this is the relationship he has with almost everyone at the CEPH. Second, there is a more complicated relationship of mutual utility and/or dependence; Cohen has this relationship with very few people—currently with Chumakov. Finally, there are those who aspire to this position of equality; unless Cohen needs them badly, such a move leads to a rupture, as with Lathrop and Froguel.

The Spy Who Came in from the Cold

Waiting in the hallway as I go in to see Cohen is a youthful man, perhaps in his early thirties, in a dark gray suit. Cohen keeps him waiting for close to an hour. He introduces himself: his name is Laurent Alexandre—he is a surgeon, a polytechnician, and an "Enarque" (Ecole Nationale d'Administration), as well as an economist! He describes himself as a "fan" (in English) of Cohen who has come to offer his services. He works in the government and has some suggestions on how to organize the politics and lobbying to prevent the commission organized to rule on the status of the DNA from doing great harm.

Over the next four hours, we have a wide-ranging discussion covering the most juvenile of futuristic visions: from master races being manufactured by Asian nations to dominate the Occident (Cohen at one point says, give me ten years and I could produce superstrong warriors!) to rather detailed technical proposals to show how the investment in biotechnology is profitable for the state. Alexandre insists that there is a widespread ignorance of the most elementary things about genetics in the government. It is imperative that a good deal of pedagogy be done.

Innovation: How should the decisions about biotechnical and bioethical issues be legitimated? This is a particularly troubling issue in France. There is no official body with legislative or executive authority. The legislature might well produce legislation, but it remains to be seen how the constitutional council and other such bodies would rule on it. France does not have the kind of intermediary judicial bodies that exist in the United States for gradually elaborating such matters. The state sets the law with no input from the the jurists and judges. *Hybrids:* Hybrid structures need to be created between the state and the individual, between public-supported science and industry. *Industrialization:* This is taken by INSERM types as the basic insult which Cohen is proposing to them. Both Cohen and Laurent insisted that the meaning of this word in French is broader than the English sense and amounts to "modernization." Anything that leads away from the artisanal and corporatist interests and practices which dominate in France is qualified as "industrial." This attitude works strongly against the lobby of specialists. "One must not insult the future."

APRIL 5, 1994

I have a discussion with Schachter in the morning and then spend several hours with his group, during which he gave them a summary of the research reports from the prestigious Gordon conference he had attended in the States. Then I have a several-hour lunch with Schachter and the two Belgians. A brief hello to Mariano Levin, just off the plane from Buenos Aires.

The general overview theme of the day was the necessity for a new set of paradigms or a wider and broader vision of the place of genetics in biology. This was as true for "aging" as for "life." Schachter had a discourse that was straight out of Canguilhem on the need to understand the question of aging from the perspective of normality and not simply from either pathology or in vitro artificiality. There needed to be a more complex understanding and posing of problems that could be answered. It was not so much the need for theory because in fact the field has suffered from too much theory and too little experimentation or "hard facts." Rather, there was what he calls a centrifugal action in which a great number of different specialists and specialties work on "aging" as a problem or question and then leave it to return to their own work. Aging is a location but not a real field. This was striking in the presentations he reported on, many of which were well done but did not connect to each other and often contradicted each other. Either there is no clearly defined object or there are false problems. Schachter dreams of going to the Santa Fe Institute and working on aging.

Telerman and Amson's two main topics are (1) their view of Cohen and the CEPH and (2) Life. The latter is in a sense simpler. For them, some central mechanism must be involved in the life and death of an organism. They used the metaphor of the inner sanctum, the holiest of holies, in which the mechanism of life or death is to be found—likely to be a small number of genes. This long discussion combining cDNA display and other technologies with Hebraic mysticism or vitalism hovers between the sublime and the ridiculous.

They have an equally starry-eyed vision of Cohen. He is the visionary of the biosciences. He is one of only a very few people they have presented their project to who listened and understood what they were up to. Their view of the CEPH as a technological Mecca seems amazingly naïve. They express no reservations about the CEPH.

The central theme of the day is flexibility, variability, and difference as the norm of health. Schachter gives me an article on heartbeats that shows that the normal beat is widely variant and exhibits a great deal of microfluctuation; the variation in heartbeats is reduced to a standard "normal" curve in old age and indicates a decreasing flexibility to respond and adapt.

Moncany shares this CEPH-as-refuge vision. He was "taken in" by Cohen and given only the encouragement to think (so far no lab space). Perhaps he is naïve in his judgments of people, but he errs on the side of generosity and trust. Moncany and Telerman (Schachter worked in Gallo's lab) have played in the elite labs and are now pulling away from the institutional side to pursue "science as a vocation." They have a "higher calling" not only for the "good of humanity" but also for the pursuit of the truth. This is the spirit of Dausset as continued, modernized, and even industrialized by Cohen.

Cohen's remark that the Americans at Millennium took care of both the business and the legal side of things and left him to think about the vision of the scientific project is telling and consistent with this picture. He was clearly irritated by the French financiers, who didn't offer to provide this service.

APRIL 11, 1994

Laurent Alexandre arrives and we spend four or five hours in discussion. His view of difference is on two axes: civilizational (Islam, Asian-Confucian, and Occident-Lumières) and biogenetic (genes for aggression). He retreats on these to some extent under questioning but these are his folk categories. He is haunted by the oppression of women in Islam and even more by the possibility that the Asians are unconcerned about the eugenics threat and might create a race of Rambos. He is not a partisan of the Universal Rights of Man arguments. He prefers Nietzsche and Clausewitz. There is no way to integrate the Rights of Man documents with the "plurality of genomes." It is not genetic diversity that troubles him but social or communitarian pluralism. The Americans have missed the boat. This man is the product of the most elite of France's schools. What is going on?

Alexandre, over lunch, tells me that he has been sent by the government to check up on the CEPH. He has concluded that I am a spy for the Americans. "Which Americans?" I ask. "The government," he answers. "That makes twenty-five years," I tell him, "since I was last accused of being a spy. It was in a Moroccan village. Several villagers told me point-blank they had figured out that I was a spy for Israel." I told him this reassured me that I was doing fieldwork. We finished lunch. He did not return.

APRIL 28, 1994

Cohen needs to talk. He tells me "in strict confidence" that many people are approaching him with job offers, several in the United States (American, multinational, and French). Although he is continuing the discussions, he cannot imagine living in an American suburb. If he left, where would that leave the CEPH? As a service institution providing family DNA or YACs. Dausset could finish his career in the state of purity he so cherishes.

Cohen wants to take the next step into the major leagues of entrepreneurial molecular biology. He is adamant that the step to multifactorial issues is a megastep, not a simple addition to things as they are now. It will require large amounts of money and organization. He expresses mixed feelings about Millennium. Millennium had been criticized for not providing enough money for the diabetes project, but subsequently, when they made the alliance with Hoffmann La Roche, it became clear that the financing offered to Millennium could never be matched by the resources that the CEPH or the French government had at its disposal. While funding is a challenge, if he is to stay in France, he needs to derail the nationalism issue by finding a French connection.

May 8, 1994

Beckmann tells me that Froguel was told that the Cochin Hospital was not going to give him the laboratory space they had promised him. The other shoe has dropped. Apparently there will be no space in Paris for Froguel.

The CEPH Scientific Council

As mandated each year, the CEPH has constituted a scientific council composed of French and international scientists, who hold an annual evaluation meeting. The council has received a precirculated report (assembled by Rebollo), and the day's proceedings were little more than a reiteration of what was in the report. Cohen simply underlined the past triumphs and current strengths of the CEPH. He left out entirely or glossed over quickly much of its current activity (aging, AIDS, cancer).

Cohen states that the problems with Froguel are the result of a personality conflict; the CEPH is supporting him and hoping to find him space elsewhere (but the project is out of date in any case). Questions are raised about the status of the other researchers in Froguel's group. Cohen avoids giving a definitive answer.

The final report is hashed out in less than an hour. It is strongly positive. The report stresses the international importance of the CEPH's achievements in mapping and its glorious record of international sharing of materials and data. Some doubt is cast on the viability of the candidate-gene approach. It is recom mended that the cancer and aging projects be reevaluated in a year. In conclusion, the report praises the CEPH for placing France, *notre pays,* in such an excellent international light.

May 25, 1994

Froguel's situation is worse and worse. Oxford is delayed at least a year. He has possibilities in Strasbourg, Lille, and Montpellier. This delay is bad for Froguel and bad for the CEPH. Everyone is waiting. Where is all the nationalistic fervor now? French DNA has no home.

Resolution but No Solution

The Government Commission
The government commission did produce a report. It had no apparent impact. Attempting to find someone who had read it has proved difficult. Four years later, as this book was being written, obtaining a copy or locating the one that was presumably sent to the CEPH has proved impossible. After the government changed in spring 1997, there was a reshuffling of personnel, and the new officials have no knowledge of the report.

The Nouveau Marché
The government did introduce institutional changes in the French stock market system. A government study group started work in January 1995 on innovations that would stimulate and aid activity in the high-technology sector. French government efforts to support large-scale computer companies had been a catastrophe. Although there were a certain number of biotech companies in France, and a few successful ones, their numbers were minuscule compared to the American situation. Among the many reasons for this state of affairs was the fact that companies that had not been making money could not be listed on the French stock exchange. Most biotech companies lose large amounts of money for a relatively long period of time. Further, French tax law did not

provide tax "write-off" provisions that would encourage capital to be invested in start-up companies.

The government commission advised establishing a subsidiary stock exchange with different rules of entry as a means of supporting the growth of high-technology companies in France. The French government created the Nouveau Marché in February 1996 as a subsidiary of the Paris stock exchange. The first public stock offering was made in March 1996. In the first year four were listed. The Nouveau Marché has no minimum requirements on company age, trading history, or the percentage of company equity to be offered. Stock offerings can be listed in any currency and there are no profitability requirements. A minimum-issue volume of $2 million and earnings of at least $1.6 million a year in equities are required. The Nouveau Marché works like the American NASDAQ in that a sponsor is needed for entry. Candidates are then evaluated on future development and growth prospects, in contrast to traditional European standards.

Initial reaction from the financial press was cautiously positive. A *Business Week* article coyly titled "Come Home, Little Startups, Europe Wants Its Own NASDAQ to Capture More IPOs,"[1] opened by observing that since European investors have long shunned homegrown technology start-ups, they have had to go to NASDAQ to raise cash. European governments, including the French, are making efforts to keep this capital at home. It will not be easy to compete with the Americans in listing the best continental tech companies with global potential. NASDAQ listings sometimes claim valuations that are five times higher than similar listings in Europe. The continent also lacks a research community that can analyze cutting-edge products and manufacturers. The article observes that this attempt is not "the first time Europeans have tried to create a venture capital culture. An effort in the mid-1980s went down in flames when a host of funds and investors lost millions." This time officials promised a more careful approach. *Business Week* was cautious, even patronizing, wondering "will Europe's new markets have the liquidity to succeed? Europe had only 24 IPOs last year." The Europeans insist that this time they are ready to succeed—"but no one's handing them a victory yet."[2]

Genset

Almost immediately after the collaboration with Millennium started to run into trouble, Cohen began to look for alternatives. He never waivered in his analysis of the situation: French government funding was neither sufficiently ample nor sufficiently flexible to provide the basis for success in the world of biotechnology. Cohen explored multiple options, including lucrative offers in the United States, alliances with large French pharmaceuticals, and no doubt other venues. He decided to enter into a relationship with a French company named Genset. This decision led to extremely bitter negotiations between Genset and the CEPH over the conditions of departure of Cohen and the core genome mapping team at the CEPH (some thirty people). After more than a year of haggling, Cohen and his team left for Genset. The terms at the end of the negotiations were not substantially different than at the beginning. Much had been destroyed in the process. Central and most symbolically important: the relationship between Cohen and Dausset deteriorated badly. By the time of Cohen's departure, discussions were conducted in a hateful and hurtful manner. A close bond and a productive partnership were shattered.

The Febuary 8, 1996, issue of *Nature* published an article titled "French Genome Pioneer Goes Private." Genset plans to develop a comprehensive approach to genomics ranging from research to drugs with the arrival of Cohen. Genset will use gene identification to develop small-molecule therapeutics.[3]

The Wall Street Journal, March 1, 1996, ran an article titled, "With Arrival of Research Star, Genset Embarks on Hunt." Daniel Cohen will join Genset in March 1996. Genset had manufactured synthetic DNA. With Cohen the company hopes to track down "susceptibility" genes for major diseases and then develop drugs and screen for effectiveness. Genset becomes one of the first European entries into a field thus far dominated by U.S. biotech companies such as Millennium Pharmaceuticals, Inc. To catch up, Genset promises economies of scale to gene hunting. "It's a brute force approach," Cohen says, "and we guarantee to find any gene for a common disease in less than two years." That's an audacious claim—considering that most successful gene hunts have taken six to eight years. But Cohen has a proven track record in big-ticket research projects. Initially Genset's gene hunting will be limited to three diseases—prostate cancer, schizophrenia, and osteoporosis. Someday, Genset chairman Pascal Brandys envisions using the chemistry prowess honed through synthetic DNA production to develop candidate drugs.[4]

In June 1996, Genset raised close to $100 million in an initial public offering on NASDAQ. Genset is also listed on the Nouveau Marché.

Genset's Web site reads (in part):

> Genset is engaged in the systematic and comprehensive analysis of the human genome to identify and patent genes and regulatory regions related to selected common diseases. Genset's objective is to apply its genomics technology to discover drugs for such diseases and to enter into strategic partnerships with pharmaceutical companies which will develop and market these drugs. Genset applies large-scale industrial techniques and has assembled an integrated array of technologies comprising high resolution mapping, sequencing, NetGene(r) database and BioIntelligence analysis software, functional polymorphism scanning and DNA synthesis, to accelerate the identification of disease-related genes. Genset's initial commercial strategy is to target prostate cancer, schizophrenia, osteoporosis, selected dermatological diseases and cardio-vascular diseases. In May 1996 Genset has entered into a strategic alliance with the French pharmaceutical company Synthelabo to discover genes associated with prostate cancer. Seven months later Genset announced that they have identified several chromosomal regions containing suspected genes associated with prostate cancer. In September 1996, Genset entered into a research collaboration with Johnson & Johnson's wholly-owned subsidiary Janssen Pharmaceutica for the discovery of genes involved in schizophrenia.

Return of the Prodigal Sons

During the second half of 1997, it is announced that both Mark Lathrop and Philippe Froguel, who had accepted the position of director of the human genetics department at the Pasteur Institute

of Lille, will return to the Paris region, with offers to head parts
of a government-supported genome mapping and gene location
project that will essentially replace Généthon in its mapping tasks.
Friction ensues and the offer to Froguel does not work out. The
AFM is actively exploring possible gene therapy venues. It is close
to opening a new venture, Généthon III, for gene therapy but
decides to wait until the technologies are improved.

France Honors Science
In 1997, the French government awards Daniel Cohen the Legion
of Honor.

NORMALIZATION

The moral of our story, at one level, is simply one of a *retour à
l'ordre,* the service, Michel de Certeau proposed, that the purgatorial machine renders to the reigning powers. These powers have
successfully reasserted their prerogatives: the intrusion of things
identified as foreign has been stopped at the borders; ethical rigor
has been announced; some internal institutional realignment has
been grudgingly granted. All in all, the yield is a perfectly respectable, if minimally innovative, arrangement for official French institutions. After all, Genset has been very astute in raising money,
the French state has maintained an official presence in the genomics world, and Froguel will apparently return to Paris (to the
hospital Hôtel Dieu) with his DNA of French diabetics. Although
it is true that the CEPH has lost its former luster and creative
energy, the decline of a once innovative institution is to be expected in the order of things.

Still, although things are functioning normally in the purgatory machine, all is not well. By chance, as I was writing this
conclusion, *Science* published a News and Comment piece, "Has
French AIDS Research Stumbled?" "Once, French scientists were
at the forefront of AIDS research; now they are struggling to keep
up. Most say the country's hierarchical research system is the cul-
prit."[5] The piece rounds up the usual journalistic commonplaces.
In a different genre, it too is a kind of call to (a more modernized)

order. French scientists willing to be quoted in the piece pointed to a number of factors in the loss of France's initial predominance in AIDS research. Although France has proportionally the largest AIDS research budget in Europe, it is outspent by the United States in absolute terms by more than 30 to 1. Where the outlays of multinational pharmaceutical firms figure in these nationalistic calculations is not made clear. Still, French scientists told *Science*'s journalist that there was more than money involved; the problem was more deeply rooted in French soil. France's hierarchical research structure lacks dynamism and flexibility. It "stifles creativity, rewards mediocrity, and—the greatest sin of all—places serious obstacles in the way of young scientists seeking an independent career." One obvious step toward a solution would be the creation of more postdoc opportunities. However, the source of the problem, some opined, was deeper yet. It was cultural, a question of values. One researcher observed, "We [French] are not opportunistic." *Science* enthusiastically took up this comment, glossing it as follows: "The relatively unaggressive nature of French science may be the key to understanding why France is lagging. . . . Those with less desire to win are going to lose." In its own cultural blindness, this claim confounds "aggression" with entrepreneurial competitiveness, but in *Science*'s own economy of motives and causes the argument had been made.

What is to be done? Willy Rosenbaum, one of the first clinicians to work on AIDS (one imagines him responding with a certain world-weary Gallic shrug and a soulful drag on his cigarette), proclaims, "The only hope for the future would be complete reconstruction of French research institutions." Not everyone in France is this despairing; the article concludes on a more upbeat note: "Who knows?" another leading researcher wistfully declared, "perhaps right now there is a lab in France writing up a paper that will revolutionize everything." Hope springs eternal: if France lacks dynamic institutions and sufficiently modern values, then at least it is still possible that it has ideas. Science, *Science* would no doubt concur, retains a certain autonomy from culture, institutions, economics, and politics both local and global. For those looking for an explanation, the 30 to 1 spending ratio seems more significant than the lack of aggression among the practitioners of French science. Still, something more *is* at stake in both the AIDS and the genomics stories.

"Tensions Grow over Access to DNA Bank" proclaims the headline of an article in the News section of *Nature* for February 19, 1998.[6] The article reports on a controversy over an arrangement between the CEPH and Genset for access to the data banks of Project Chronos. The controversy was triggered by the dismissal (or procedures to dismiss) François Schachter from the CEPH. Schachter had been instrumental in setting the bank up in 1991. "Schachter, who plans to challenge his dismissal in an industrial tribunal, is expected to argue that his sacking stemmed from his opposition to the Genset deal and the way in which confidentiality clauses in the agreement hampered his collaboration with groups elsewhere, and that his eviction from CEPH denied him access to his research materials." The CEPH denies these charges. "Gilles Thomas, who was appointed scientific director of CEPH in 1996, says the sacking was unavoidable, alleging that Schachter refused to cooperate with management and infringed confidentiality agreements." Under a 1996 contract, Genset had first-refusal rights on results from Chronos and other CEPH research into aging for three years in return for Genset's funding the program at $5.3 million. The contract also forbids the CEPH from passing information obtained from its research to a third party without Genset's authorization. This clause includes oral presentations at conferences.

"In a separate development, CEPH is embroiled in a legal dispute with Genset about what it claims is the latter's refusal to pay installments of the 32 million francs to CEPH as agreed in the contract."

Axel Kahn, a member of the original government commission investigating the CEPH-Millennium situation, now deputy scientific director for life sciences at the French pharmaceutical giant Rhône-Poulenc, asserted that although the commission's report was never translated into law, the manner of Schachter's ouster runs counter to its spirit and that it was "obvious" that researchers had a right to exploit their own material. Kahn was critical of the CEPH-Genset agreement: "A moral engagement was made with the people sampled, which demands that they be informed of the commercial exploitation of their genes." A representative of the Ibsen Foundation said he felt "betrayed" because the centenarians whose DNA was used were never told it would be patented.

Schachter has received backing from colleagues. National coordinators of a European Union network on "molecular gerontology" wrote to Jean Dausset to express their concern. They worried that "a basic principle of our scientific community, freedom of research, seems to have been compromised." Other colleagues said that France had taken the lead in such investigations but "the country's reputation is now being damaged."

"What Ails French Biosciences?" asks a News and Comment piece in the March 6, 1998, issue of *Science*. The piece reports on a meeting of top biomedical researchers in Paris, who came together to discuss new reform plans set forth by France's research minister, Claude Allègre, a geochemist.[7] Allègre has had a controversial reign, provoking the ire of various parts of France's research establishment with critical comments and proposals to streamline procedures he has made since his appointment, after the Socialist electoral victory. One of Allègre's key points is a nationalist one: French research should serve France's national interests. The article highlights a quote from the minister: "Reforms are needed because 'Europe is being eaten alive by American industries.'" At a conference held to address the issues there was a good deal of mixed reactions to the minister's proposed reforms. However, a consensus existed that "France, despite its long and proud history in biomedical science, has fallen considerably behind many other countries in making this research pay off in economic terms." National Assembly President and former (Socialist) Prime Minister Laurent Fabius, in his opening address to the conference, said, "As for our biotech companies, their situation is often disturbing, if not downright bad." France's biotech sector was generating less money than that of Israel. Another official remarked, "We cannot remain in such an underdeveloped state."

Philippe Froguel, a chief organizer of the conference, criticized the lack of a patent policy that would lead to commercialization by France's leading research institutions. He also criticized "the conformism in the academic milieu that points the finger at researchers who enter into contracts with industry and accuses them of making a personal profit from research supported in part by public money." Froguel warned against the American model in which young researchers come together and "keep their eyes riveted on stock prices on NASDAQ." Rather, the main priority should be to "give more muscle to public research," so that there can be "desirable" contact with industry.

EPILOGUE

The Anthropological Contemporary

He must make things felt, touched, make his inventions heard; if what he brings back from there has a form, he gives a form; if it is unformed, he gives it no form. Find a language.

Extravagance becomes the norm, absorbed by all, it would truly be a multiplier of progress.

While waiting, let us ask again of the poets the new—ideas and forms.

All the clever ones thinking it is easy to respond to that request. That is not the case!

ARTHUR RIMBAUD, letter to Paul Demeny, May 15, 1871

What things are contemporary? Consider a late-model car. It is a disparate aggregate of scientific and technical solutions dating from different periods. One can date it component by component: this part was invented at the turn of the century, another, ten years ago, and Carnot's cycle is almost two hundred years old. . . . The ensemble is only contemporary by assemblage, by its design, its finish, sometimes only by the slickness of the advertising surrounding it.

MICHEL SERRES, in Michel Serres and Bruno Latour, *Conversations on Science, Culture, and Time*

Subject: Conceptual Activism

Daniel Cohen and I met at a panel on science and society at the science museum at La Villette in 1992. We hit it off: I was enthusiastic about what he was doing, I had criticized the right people (his enemies in the genome world), and, as the saying goes, the meeting had the right "chemistry." (No one yet claimed to have found the gene for friendship.) He took time from an extremely busy schedule to give me a tour of his lab in Paris as well as Généthon. When Cohen invited me to come to the CEPH, he qualified his invitation in an intriguing fashion. He insisted that while he did not want an "ethics committee" at the CEPH, he would welcome "a philosophic observer." The formula was pleasing, if somewhat obscure. In response to my query about what was wrong with ethics committees, Cohen gave an irritated and dismissive wave of his hand. As to what a philosophic observer might be—that, he said, was up to me to invent with them.

Cohen is not universally loved in Paris—but who is? A French friend, a knowledgeable and cautious historian of the biosciences, told me that the book Cohen had just published "brushes the boundaries of eugenics." I registered the warning but am enough of a Paris veteran not to take any such comment at face value. The book, *Les gènes de l'espoir* (Genes of hope), is an ebullient one, promoting the new genetics and molecular biology to the French public. In the book, Cohen defiantly waves the banner of progress once held aloft by Maupertuis. Cohen mentions me in the book. Writing about the CEPH and its strategy to distribute its family material internationally, Cohen wrote: "What incited us to follow that path? Something that today interests a famous American anthropologist and philosopher, Paul Rabinow, who defends the idea that such a system could only have developed in France, the country that made the revolution of 1789. He wanted to come to the CEPH at all costs to study what he calls 'the French spirit' as the main reason for our success. A decidedly Francophile American."[1] Although I doubt I ever mentioned 1789 and am sure I never talked about *l'esprit français,* Cohen was correct that I had wanted to study the CEPH because I believed it was doing something distinctive in its alliance with the AFM and that obviously the specific trajectory of the history of public health and science in France had something to do with that. Regardless, Cohen's book had been drafted to serve his own purposes. Consequently, I entered into the fieldwork at the CEPH in a relationship of complicity, in a rather negative sense for some in Paris (the book was published before my arrival) and a rather more positive one inside the CEPH. I was perfectly willing to assume that relationship, as Christian Rebollo would have put it, as the price of entry. This subject position, however, was not the standard one for an anthropologist to adopt. What is one to make of it?

In an article titled "The Uses of Complicity in the Changing Mise-en-Scène of Anthropological Fieldwork," George Marcus identifies "rapport" (between anthropologist and native informant) as the basic figure of classic ethnographic practice and then sketches the conditions that have occasioned its displacement. From Bronislaw Malinowski through Clifford Geertz, Anglo-American anthropology put forward an intensive, trusting relationship with one or several informants as the heart of its methodology and built its authority upon that relationship. During the last ten years or so, the critique of that figure of "ethnographic authority" (again, in the Anglo-American academy) has been massive. However, as Marcus points out, what comes after the decline of bounded cultures, empathetic fieldworkers, wise natives, and backgrounded power apparatuses is far from clear epistemologically, ethically, or politically. Regardless, Marcus underscores a key point, "the recognition of ethnographers as ever-present markers of 'outsiderness.' . . . It is only in an anthropologist-informant situation in which the outsiderness is never elided and is indeed the basis of an affinity between ethnographer and subject that the reigning traditional ideology of fieldwork can shift to reflect the changing conditions of research."[2] In my case, the ridiculousness of the spy charge leveled at me by the French government emissary lay in its outlandish literalism; the questioning edge of the journalist's assignation was simply too eager; the cast of Cohen's description of his invitation to go to the CEPH too unabashedly stereotypical. Collectively, however, they square the circle; the anthropologist was a conduit of potent forces, if not their representative. The American challenge, *le défi américain,* was once again present in France. That positioning provided the basis for an excellent rapport within the CEPH. The rapid affinity with

Cohen, Froguel, Rebollo, Schachter, Telerman, Moncany, Levin (or for that matter with Alexandre and Bensimon), and a series of others was structural and conjunctural.

Marcus perceptively observes that the anthropologist, "by virtue of these changing circumstances of research, is always on the verge of activism, of negotiating some kind of involvement, beyond the distanced role of ethnographer, according to personal commitments that may or may not predate the project."[3] Cohen's choice and his open-ended demand to me amounted more or less to this: do not arrive with an a priori ethical armamentarium but work with us to develop an informed experimental and reflective practice linking genomics and social responsibility. The demand was for a kind of situated conceptual activism (without a Cause). It permitted a relatively knowledgeable (about France, about the United States, about genomics, about biotechnology companies, about scientific journalists, etc.) figure the possibility of refracting pressing decisions being made at the CEPH about advancing (or retarding) action. The refraction would be through a kind of engaged facilitation rather than through a process of external judgment or discursive clarification in a foreign technical discourse such as philosophy or ethnography.

Among other thorny problems, the invention of such a subject position certainly presents a challenge as to how one can incorporate a sense of "disinterestedness," one of the primary diacritics of science, whether human or natural. However, it presents it in a manner that includes both the scientists and the ethnographer, even if asymmetrically. The exercise required to achieve this state is habitual in the natural sciences, but that very ingrained disposition poses the problem of establishing limits to an uninhibited scientific curiosity that cannot legislate for itself according to its own vocational canons. Disinterestedness has come to have negative connotations in certain factions of the human sciences. However, the justification for some of this negativity is superfluous, as by "disinterestedness" one could simply mean a certain prima facie attentiveness to the way things are and an alertness in the face of that situation. This disposition need not entail an acquiescence or any dampening of emotion. It does mean a certain vocational integrity, an asceticism in Weber's sense, but also in Foucault's, that is to say, a certain rigor and patience that could, if we are fortunate, lead us somewhere else, beyond what we already believe and know.

Finally, to continue Marcus's conceit, such a subject position poses the status of the contemporary in a fashion that entails a certain affinity, a shared concern, a willingness to enter into potentially compromising situations, temperaments that risk engagement with a certain disinterestedness, a certain distance, so as to get closer to things. We all shared a sense that the question "what is to be done?"—with biological research, materials, and vocations—could not be evaded. It could not be evaded because it needed to be addressed per se. However, it also needed to be confronted precisely because powerful, well-funded, and/or richly endowed with cultural capital others were in the process of answering it. The pressure of events yielded an affinity etched with anxiety, even if the anxiety of each party was not precisely about the same things. The relationship was not characterized by irony, the master trope of modernism. Nor was it marked by "free play," the dogma of postmodernism. Classical ethnographic distance was not the position adopted, although that hardly means that one's critical faculties were left in the library; the jokes about "he is studying us under a microscope" stopped after a few weeks. Nor was the anxious affinity marked by a confessional practice of autobiographical reflection or simple construction of the other. We were operating in a situation that was, to a degree, uncharted or experimental, which is not to say that we had no conceptual or existential resources to rely on.

METHOD: ANTHROPOLOGY OF
THE CONTEMPORARY

French DNA is not intended as an example of "the history of the present." The phrase (yet again) is Foucault's although he had very little to say directly about what he considered the history of the present to be. Minimally, one can say that it is (a) diagnostic of a current problem, (b) primarily genealogical in its elaboration, (c) not focused in its substantive discussions on contemporary instances. The history of the present is neither properly historical nor sociological nor ethnographic. Thus, *Discipline and Punish,*

Foucault's "prison book," and one in which the term "history of the present" actually appears, arose (in part) out of a concern with prison uprisings in France during the early 1970s.[4] The bulk of the book treats issues and events located primarily in the eighteenth and the first half of the nineteenth century. Foucault was clearer later on that he was not talking about a "carceral" society that encompassed all the practices of the day in an exhaustive manner. He had not written a history of the prison but analyzed the genealogy of elements of a disciplinary technology. It is fair to characterize Foucault's subsequent work as elaborating a more complex understanding of the history of the present, although it is really only in his interviews that he talks directly about what he considers to be the salient problems of that present.

My own *French Modern* sought to be a history of the present. It began with what it diagnosed to be a crisis in social planning and its place in the welfare state. It provided a genealogy of a number of the elements that had been gradually assembled (over the course of two centuries) and stabilized for a time in the practice of urban and social planning. It sought to describe a certain political rationality embedded in a form of governmentality. Although it drew part of its impetus from ethnographic fieldwork in Morocco and long experience in France, *French Modern* was certainly not a standard ethnography. Among other things, it was not written in what anthropologists used to call "the ethnographic present." Nor did it emphasize the mode of subjectivation of the inquirer; rather, it was devoted to assembling the elements of an object. In that sense, it was an analysis of the long-term emergence and articulation of a form once regnant, now in decline. Finally, it sought to be inventive in the shape it gave to its narration.

Neither *French Modern* nor *French DNA* are hermeneutical books in a standard sense. Although the word has many meanings, two can be eliminated at the outset. The first is "commentary" or "thick description." The origins of modern hermeneutics lie in biblical criticism, especially the nineteenth-century historicization of the Bible and then other authoritative texts. In both its biblical and its secular form, the quest was to recover the original meanings carried by the text (or intertext). Such commentary has a long and distinguished lineage; its culmination was attained in the human sciences in the textual metaphor that came to dominate sectors of anthropology and other human sciences. Such an approach has been in crisis in anthropology at least since James Clifford and George Marcus's *Writing Culture,* although new (counter) variants abound in both deconstructive criticism and cultural studies insofar as the text is still taken to be the primary locus of interest. In this instance, I might have provided more detailed textual interpretation of the documents produced by the National Ethics Committee relating them in detail to debates about the sacred or the person. There is ample material available to be commented on and many links could be established with authoritative texts from the past. I also could have taken an "'anthropological" approach to the CEPH in which I analyzed it as an organized site whose culture could be thickly described. Or I could have shown that its culture was not bounded and holistic but needed a thick description that showed how the global and the local formed a permeable, even leaky, nexus of signification. There is much to learn from such approaches, but they are not the practice that I am engaged in.[5]

The second meaning that can be eliminated is the "hermeneutics of suspicion." Paul Ricoeur coined this telling appellation to characterize an approach to understanding shared by Marx, Nietzsche, and Freud.[6] Ideally, the position asserts that the task of analysis is to uncover an underlying disguised truth. That act requires an authority who (after rigorous self-analysis) has already seen the truth or the path toward it and is able to guide the self-deluded participant through a series of steps and so lead him to also see what had been obscured. Individual insight, however, is neither sufficient nor trustworthy. In order for such insight to be effective, for it to change the subject, there has to be an acknowledgment, a form of confession to the authority figure. Finally, the goal and payoff of this asceticism, purification, remaking, is a kind of liberation, or at least setting the subject in process toward that goal. Variants of this practice continue to flourish in American Lacanian circles, especially in literature and film studies. The French bioethics discourse is often cast as a hermeneutics of suspicion as well. One might have joined such an endeavor by highlighting the will to power omnipresent in the leading actors, or their gender politics, as the shaping force of how modern science constitutes its social relations as well as its relation to nature, or by simply highlighting the insidious and perverting penetration of knowledge by capital and the state, and thereby uncover, reveal, and demand a recognition of these forces and institutions as what was really at issue. By so doing, one would have constructed oneself as on the path to liberation. Again, I am not denying that such forces are there nor that they are important. However, I was (and am) attempting to do something different.

What then am I trying to do? In taking up a research project, one can choose to emphasize that aspect of things that "tries to insert movements, figures, stories, activities into some larger organization that predates and survives them."[7] Standard historical or sociological accounts usually fall into this genre but so too does the history of the present even if its goal is to disrupt that organization. In contrast, one can privilege the attempt to "release figures or movements from any such organization that predates them and survives them, allowing them to go off in unexpected directions or relate to one another in undetermined ways." Of course, these two figures are not exclusive and privileging one does not mean being blind to the other. This method seeks to identify (following Gilles Deleuze) "those forces or potentials whose origins and outcomes cannot be specified independently of the open and necessarily incomplete series of their actualization. Such is their multiplicity that it can never be reduced to a set of discrete elements or to the different parts of a closed or organic whole."[8] Hence, one is seeking to identify neither the instance of a generality nor the concretization of an abstraction nor some mysterious thing in cyberspace or "glocal" (i.e., global/local). Rather, one seeks "as yet unspecified singularities," assembled in action. An experimental mode of inquiry is one where one confronts a problem whose answer is not known in advance rather than already having answers and then seeking a problem.

OBJECT: FORMS/EVENTS

Another salient aspect of theory and practice is worth underlining. *French DNA* is not about Culture in the sense that Clifford Geertz, Marshall Sahlins, or Claude Lévi-Strauss have so brilliantly developed it. Nor is it about Society as Pierre Bourdieu or Anthony

Giddens or Jurgen Habermas have so profoundly theorized it. *French DNA* contains no totalities, formal systems, encompassing fields, epochs, worldviews, universal subjects. Nor does it even contain Theory in the traditional sense of the term that would make the empirical material into a case study, an example, a testing ground. It is not about Globalization or Postmodernity or the Background Practices.[9] So what is it about? In order to begin to answer that question, I turn (again) to the work of Michel Foucault: "It is easily believed that a culture is more attached to its values than its forms. The latter, it seems, can be readily modified, abandoned, reworked: only meaning is profoundly anchored. This is to be mistaken about how much forms, when they unravel or at the moment of their birth, can provoke astonishment or arouse hatred."[10] Foucault was discussing avant-garde music, but the distinction is illuminating as well for the story at hand. The form created in the United States called (somewhat imprecisely) the biotechnology industry has aroused a riptide of affect in France that goes well beyond Nietzsche's diagnosis of *ressentiment* and the will to "spiritual revenge'" as the characteristic malady of thinkers (although it contains elements of that affect as well). The reaction is linked to a combat of forms. As an ethnographer, who had accepted Cohen's challenge to be a "philosophic observer," my task was to identify the crystallization of value judgments around new forms, not to adjudicate those disputes.

The CEPH and the AFM, both individually and together, invented one of the most axial and edifying forms seen since the (recent) appearance of genomics. It lasted a number of years and was dissolved—a possibility envisioned, even mandated, at the outset. The form provided for a productive, synergistic juxtaposition of an unorthodox technoscientific program and a normativity arising from the demand by patients groups for a scientific and therapeutic return on their investment, or, at the very least, a serious recognition, in the way research was organized, of their perilous state. Cohen and Barataud—restless and curious, neither comfortable with the ironic Nietzschean view that we have all the time in the world to experiment with ourselves yet among this view's most dynamic "specific intellectuals" of the new genomics, in a strong sense—needed and found each other. Barataud needed (although it took him time to understand and realize it) a restless cosmopolitan capable of assembling (from insights drawn from around the world) a program of industrial innovation of the technologies of life. Cohen needed a source of funding that would free him from the constraints of France's *noblesse d'état* (France's reigning scientific and medical mandarinate) while providing him with a source of social legitimation, which he both needed and desired. The dependency on a network of activist patients whose keen present suffering and fateful bodily degeneration lent not only an urgency but an ethical imperative—and accountability—to his own relentless ambition and curiosity. Finally, the time was propitious. Others have felt the need to advance a medical or scientific cause just as keenly; other productive partnerships have been formed in many areas, including genomics. However, the few years this partnership endured were pivotal ones in genomics. The CEPH-AFM collaboration emerged in its distinctiveness during this singular conjuncture.

The form/event created by the CEPH-AFM collaboration was, as I have said, an impressive, initial attempt within the rationality of genomics to meld *zoē* and *bios,* to bring the sheer genetic constituents assembled in yeast artificial chromosomes and computer data banks into a relationship with a form of life that recognizes the contemporaneity of this knowledge and attempts to shape it accordingly—and it aroused both surprise and suspicion. The French public has continued to respond with an emotional and monetary ardor to the Telethon. An unexpected and unpredicted space was created for public expression and agenda setting that could be taken as an admonishment to the state for its lack of response to pressing health problems and that was as well the affirmative creation of a public sphere for citizen participation. The form that sphere has taken is not without its ambiguities and dangers, and much criticism has been leveled at it. In some quarters, and for varying reasons, an ethos of wary vigilance reigns among those who feel that the CEPH-AFM partnership perverted the process of state regulation of bioscence and medicine and thereby created a "democratic deficit." At one extreme, as we have seen, conferences brought together the most prestigious authorities in Paris to consider, with that degree of high seriousness Sartre had mocked as the bourgeoisie's "l'esprit de sérieux," proposing ritualistic safeguards before approaching the human genome. Although these gestures might seem exotic to an American audience, they should be taken as a traditional French reenactment of the sacralization of society as it stood. They are a reactive call to symbolic order in the face of forces in the present that are being pried loose from their previous webs of social practice and representation. At another level, leaving the funding of long-term biomedical projects (the myopathies but also AIDS) to the vagaries of the media and thereby leaving them vulnerable to changes in fashion or mood as well as scandals is problematic and is unlikely to endure.

The CEPH-AFM collaboration was not the only original form invented in France during the 1980s. The French National Ethics Committee, as its proponents underscore, is unique in the world. In a spirit of high seriousness, the committee took up its distinctive mandate to create a site of national debate and reflection, carrying it forward at a deliberate tempo. The committee has separated itself just enough from legislative and legal institutions to not appear partisan, yet it has remained close enough to those institutions to set an agenda. It has skirted the role of authoritative producer of truth although its members are prolix about articulating what they take to be fundamental principles. The committee has served, as intended, as a site to clarify those principles: "the human body is not a subject of commerce" (blood, that is—not placentas). Perhaps its singularity lies in its creation of a forum to disseminate values and ideas. The committee's temporality has purposively been slower and more measured than that obtaining in the practices it seeks to reflect on and to regulate. It would be unseemly for the defenders of universal principles, the sacred, and the spiritual to be agitated or inconstant—yet, the defense of dignity requires an active vigilance. The future is at stake and human intercession matters.

The limitation of the committee's form in the current situation is twofold. First, by its very constitution, it is engaged in a critical practice in the Kantian sense of limit setting. As concerns retrospective matters, such critical practice is appropriate, even vital. An alert and insistent vigilance is entirely appropriate when it comes to any return to negative eugenics or violation of patients' rights in medical experimentation or an emergent biologically based slavery arising out of global commerce in bodies and body parts. That is to say, this form of ethics has an authoritative role to play when it comes to existing life-forms and known forms of life with their potentialities and dangers. Second, however, the committee's "this-worldly mysticism," its commitment to fixed (if empty) concepts of the human person, dignity, beneficence, and solidarity, risks universalizing values and ideas embedded in the past. As Weber indicated, this form of practice tends to be unwittingly oriented toward the salvation of the subject at the price of indifference to the particularities and singularities of the world. Its limits and dangers lie precisely in its inability to apprehend new things and its tendency toward pastoral normalization of those whose curiosity gets out of line.

In both the United States and France, there is a broad consensus of belief that the dangers and potentials of the life sciences and welfare institutions are identified through value conflicts and clarification. One is for or against abortion, for or against immigration, for or against the commerce in blood, for or against surrogacy, for or against the patenting of life-forms. While at one level there is much discord, at another there exists a striking consensus about what to fight about and how to orchestrate one's position and that of one's enemies. That is to say, the rhetoric of value conflict is relatively stable today. Not many people, after all, would respond in an opinion survey that they are against scientific progress, health, dignity, or human rights. However, alongside such a consensus (respect human rights, increase beneficial knowledge) or dissensus (patenting violates the sanctity of life, patenting is essential to scientific progress) and the diverse clusters of practices that seek to embody beliefs lies a more obscured terrain. This terrain is the terrain of forms/events. Forms arise from time to time that make such diffuse consensus or dissensus concrete, or may well be what produces the issue or thing about which beliefs are held in the first place. Such forms may stabilize for extended periods of time—think of the French blood donation system or the postwar American biomedical research apparatus—before being destabilized or overcome or destroyed or simply made obsolete by a multitude of possible factors from the introduction of new technologies to economic reorganization to the introduction of new beings (such as viruses or venture-capital firms). New demands arise; new problems call forth different responses. Successful responses to such problematizations—and what we mean by "successful" calls for more thought—might be called forms/events. The importance of such forms/events—and "importance" is another devilish diagnostic category—varies enormously. Analytic attention to forms/events brings us closer to the shifting practices, discursive and otherwise, as well as to the shifting configurations that both shape and are shaped by such practices.

Although one cannot say with certainty that attention to such forms/events will provide a cure for the desire for redemption, it does offer a prescription for the temporary relief from the "spiritual" demands of the day. The technology that produces the spiritual is a key element in the purgatory machine. The spiritual is diffuse, devoted to identifying what makes humanity distinctive, and organized to value the universal part of the particular in human beings. Consequently, beings, practices, and forms that do not adhere or conform to these criteria—things that are singular or differ from the reigning understanding of the human and the institutions that carry and diffuse this understanding—are excluded or reformed or ignored. At times, such policing is perfectly defensible, even laudable: combating the bringing into existence of new biological weapons, defending access to health care as a minimal condition of modern life, a prudent vigilance regarding the understanding, representation, and use of genetic information. At other times, the contemporary spiritual technology produces a disposition toward a certain self-satisfaction, leading those engaged in operating that technology to ignore new dangers (like viruses in the blood supply), to confound the national with the universal (one historical system of funding and organizing the production of knowledge is not the only defensible possibility), to employ an arguably archaic understanding of the distinctiveness of the human (hence being surprised and outraged when sheep are cloned). The spiritual technology becomes its own worst enemy, activating a dangerous machinery that forestalls or inhibits the flourishing of things, practices, and assemblages that could well enhance and abet our search for a better form of life before they are either understood or communally evaluated through experience.

Mode: Nominalist

The production of the spiritual, however, is not the only way to approach things. In my work at Cetus Corporation on the invention of the polymerase chain reaction as well as in the subsequent field project at the CEPH, I gradually came to understand that the dimension or aspect of the object of study that intrigued me was its modality. The same object can be taken up under different modalities. If the object was the "event/form," the question remains, how should an "event" be approached? Marshall Sahlins provides a classic definition that contrastively helps define my usage: "An event is not just a happening in the world: it is a *relation* between a certain happening and a given symbolic system. Meaning is realized . . . only as events of speech and action. *Event is the empirical form of the system.*"[11] The French National Ethics Committee would be pleased to agree with Sahlins and attempts to enact this understanding of event and system. My claim is that although one might well attempt to take up forms/events in their relation to some underlying semiotic system, by so doing one would fail to do justice to what is happening, because in such situations there is no stable system to be included or disrupted. A "happening in the world" is what needs to be understood. From time to time, and always in time, new forms emerge that catalyze previously existing actors, things, temporalities, or spatialities into a new mode of existence, a new assemblage, one that makes things work in a different manner and produces and instantiates new capacities. A form/event makes many other things more or less suddenly conceivable. Such happenings are not reducible to the elements involved any more than they are representative of the epoch, its instantiation. Nor are such forms/events mysterious and unanalyzable although it is hard to do so with the conceptual tools at hand. It is only that so much effort has been devoted in the name of social science to explaining away the emergence of new forms as the result of something else that we lack adequate means to conceptualize the forms/events as the curious and potent singularities that they are. To identify and analyze singularities, however, is by no means to deny their contemporary status and consequently their interconnections with many other things in the world.[12]

Events problematize classifications, practices, things. The problematization of classifications, practices, things, is an event. A sensibility of constant change, and a certain pleasure and obligation to grasp it and participate in the transformations, constitute one mode of relating to things. This sensibility takes the mode of a keen awareness that the taken-for-granted can change, that new entities appear, that our practices of making are closely linked to those entities, that we name them, that we group them, that we experiment with them, that we discover different contours when deploying questions and techniques. Such a sensibility could be called "nominalist." The term "nominalism" can also be used to characterize modes of epistemology and ethics. It is helpful to distinguish between nominalism as a claim about the nature of things (there are only particular things in the world, not natural kinds) and a nominalist sensibility that seeks to shape itself in accordance with a world experienced as contingent, malleable, and open. Such an obligation, once it becomes reflexive, becomes an ethic of experimentation. Today, it is helpful to distinguish nominalism from deconstruction, if by that one means an ethic of revealing the inherent instability of all knowledge. Nominalism certainly works against the grain of established classifications, given entities, and habitual procedures of knowing. However, it does this, not as an end-in-itself, but rather as a means of knowing things more acutely.

The present is a good time to desist from employing totalizing categories like epoch, civilization, culture, and society (or, at the very least, hesitation, scrutiny, pausing, and pondering are in order before employing them). These notions are in conceptual ruins, contributing in no small part to the disarray of the nineteenth-century disciplines like anthropology, sociology, and history that were built around them. Other forms of inquiry are under way. Science studies, for example, have been instrumental in inventing and testing new analytic categories that have proved to be powerful in the sense of extending and enlivening our capacity to understand things. Bruno Latour's articulation of "actor-networks" or "immobilized mobiles" and Hans-Jorg Rheinberger's explorations of "experimental systems" are unquestionably examples of conceptual advances. One could easily cite others.[13] In part, they are advances because they pick out things we did not have adequate means of naming before. In part, they are advances because they have found a means to avoid focusing on pseudoentities like "culture" or, for that matter, "science." I am advocating the pursuit of a larger series of limited concepts. Why? Because if, as philosophically oriented anthropologists, the goal of our labor is understanding, then our concepts and our modes of work must themselves be capable of making something new happen in a field of knowledge.

Notes

INTRODUCTION

1. D. Cohen, I. Chumakov, and J. Weissenbach, "A First-Generation Physical Map of the Human Genome," *Nature* 366 (Dec. 16, 1993). There is already
a vast literature on the various genome projects. For the CEPH, one should
begin with Bertrand Jordan, *Voyage autour du genome: Le tour du monde en 80
labos* (Paris: Les Editions INSERM, 1993); Daniel Cohen, *Les gènes de l'espoir:
A la découverte du genome humain* (Paris: Robert Laffont, 1993); Jean Dausset,
Clin d'oeil à la vie (Paris: Editions Odile Jacob, 1998); Robert Cook-Deegan, *The
Gene Wars: Science, Politics, and the Human Genome* (New York: W. W. Norton,
1994); Eric P. Hoffman, "The Evolving Genome Project: Current and Future
Impact," *American Journal of Human Genetics* 54 (1994).

2. The expression "ADN français" is contained in a letter sent to the prime
minister of France, Edouard Balladur, on March 7, 1994, by Dr. Philippe Froguel
of the CEPH. The expression was later used by others, often with a certain
embarrassment or high seriousness.

3. On the theme of French-American relations, see Richard Kuisel, *Seducing
the French: The Dilemma of Americanization* (Berkeley and Los Angeles: University of California Press, 1993); Jean-Philippe Mathy, *Extrême-Occident: French
Intellectuals and America* (Chicago: University of Chicago Press, 1993); Denis Lacorne, Jacques Rupnik, and Marie-France Toinet, eds., *The Rise and Fall of Anti-
Americanism: A Century of French Perception,* trans. Gerry Turner (New York:
St. Martin's Press, 1990).

4. Some of these changes are discussed in more detail in Paul Rabinow, *Making PCR: A Story of Biotechnology* (Chicago: University of Chicago Press, 1996);
Paul Rabinow, *Essays on the Anthropology of Reason* (Princeton: Princeton University Press, 1996).

5. Providing a gallery of these arcadian detours would require another book
in itself.

I

1. Max Weber, "Religious Rejections of the World and Their Directions,"
in *From Max Weber: Essays in Sociology,* ed. H. H. Gerth and C. Wright Mills
(New York: Oxford University Press, 1946). Quotations are from pp. 331, 355,
322.

2. Ibid., p. 333.

3. Max Weber, *The Protestant Ethic and the Spirit of Capitalism,* trans. Talcott Parsons (New York: Charles Scribner's Sons, 1958), p. 182.

4. Michel Foucault, *Discipline and Punish,* trans. Alan Sheridan (New York: Vintage Books, 1977), pp. 29–30.

5. Ferenc Feher and Agnes Heller, *Biopolitics, Public Policy and Social Welfare,* vol. 15 (Aldeshot: Avebury Publishers, 1994), p. 13.

6. Ibid., p. 11.

7. Luc Boltanski and Laurent Thévenot, *De la justification: Les économies de la grandeur* (Paris: Editions Gallimard, 1991).

8. Paul Rabinow, *Essays on the Anthropology of Reason* (Princeton: Princeton University Press, 1996).

9. Michel Foucault, *History of Sexuality,* trans. Robert Hurley (New York: Vintage Books, 1978), vol. 1, p. 143.

10. Ibid., pp. 141–42.

11. Hannah Arendt, *The Human Condition* (Chicago: University of Chicago Press, 1958), p. 97; Giorgio Agamben, *Homo sacer: Le pouvoir souverain et la vie nue* (Paris: Editions du Seuil, 1997); translated by Daniel Heller-Roazen under the title *Homo sacer: Sovereign Power and Bare Life* (Stanford: Stanford University Press, 1998). Agamben's formulation and diagnosis are far more totalizing and apocalyptic than my own. Taking the Holocaust as paradigmatic of the West, he sees current developments as putting us at risk of "an unprecedented biopolitical catastrophe" (p. 188, Eng. ed.).

12. Aristotle, *Politics* 3.5.1278b.23–31. *The Basic Works of Aristotle,* ed. Richard McKeon, *Politics,* trans. Benjamin Jowett (New York: Random House, 1941), p. 1184.

13. Aristotle, *Politics* 1.1252b.30.

14. Jewish conceptions of the purgatorial are hard to pin down and do not seem to have played a major role in Judaism. I want to thank Professor Gerald McKenney for his help on this matter.

15. The distinction between "modern" and "countermodern" as well as modernity understood as an ethos rather than an epoch is found in Michel Foucault, "What Is Enlightenment?" in *The Foucault Reader,* ed. Paul Rabinow (New York: Pantheon Books, 1984).

16. Jacques Le Goff, *La naissance du purgatoire* (Paris: Editions Gallimard, 1981); translated by Arthur Goldhammer under the title *The Birth of Purgatory* (Chicago: University of Chicago Press, 1984).

17. Michel Vovelle, *Les âmes au purgatoire, ou le travail du deuil* (Paris: Editions Gallimard, 1996), p. 275.

18. Ibid., p. 277.

19. Benjamin Nelson, The Idea of Usury: From Tribal Brotherhood to Universal Otherhood, 2d ed. (Chicago: University of Chicago Press, 1969). Quotations from pp. 111, 111–12, 113.

20. Ibid., p. 136.

21. Parallel questions are treated for France by Joan Wallach Scott, *Only Paradoxes to Offer: French Feminists and the Rights of Man* (Cambridge: Harvard University Press, 1996).

22. "[H]istoires stratifiées qui jouent les unes sur les autres; il est caractérisé

par l'ambiguïté qui permet ces jeux; il est distendu entre les programmes hétérogènes qui le circonscrivent en s'y rencontrant" (Michel de Certeau, "Poétiques d'espace," *Café,* no. 1 [1983]: 82).

23. "[C]'est un rappel à l'ordre dont l'actualité se libère" (ibid., p. 83).

2

1. Alexander Dorozynski, "Gene Mapping the Industrial Way," *Science* 256 (Apr. 24, 1992): 463.

2. See Jean Dausset, Howard Cann, Daniel Cohen, Mark Lathrop, Jean-Marc Lalouel, and Ray White, "Program Description: Centre d'Etude du Polymorphisme Humain (CEPH): Collaborative Genetic Mapping of the Human Genome," *Genomics,* no. 6 (1990).

3. There is a vast literature on these topics. I would like to thank Angela Creager for her generous discussions of these issues. An excellent overview is found in chapter 5, "Campaigns and Crystals: The War against Polio and the Political Economy of Postwar Virus Research," of her forthcoming *The Life of a Virus: Wendell Stanley, TMV, and Experimental Systems as Models in Biomedical Research, 1930–1965.*

4. Saul Benison, "The History of Polio Research in the United States: Appraisal and Lessons," in *Twentieth Century Sciences: Studies in the Biography of Ideas,* ed. Gerald Holton (New York: W. W. Norton, 1972); Jane S. Smith, *Patenting the Sun: Polio and the Salk Vaccine* (New York: Anchor Books, 1990).

5. James T. Patterson, *The Dread Disease: Cancer and Modern American Culture* (Cambridge: Harvard University Press, 1987); Patrice Pinell, *Naissance d'une fléau: Histoire de la lutte contre le cancer en France, 1890–1940* (Paris: Meraille, 1992).

6. Steven Epstein, *Impure Science: AIDS, Activism, and the Politics of Knowledge* (Berkeley and Los Angeles: University of California Press, 1996); Michel Pollak, *Les homosexuels et le SIDA: Sociologie d'une épidémie* (Paris: A. M. Metailie, 1988); Mirko D. Grmek, *Histoire du SIDA: Début et origine d'une pandémie actuelle* (Paris: Editions Payot, 1989).

7. On the AFM, see C. Barral and F. Paterson, "L'Association Française contre les Myopathies: Trajectoire d'une association d'usagers et construction associative d'une maladie," *Sciences Sociales et Santé* 12, no. 2 (1994).

8. An especially eloquent account is Alice Wexler's *Mapping Fate: A Memoir of Family, Risk, and Genetic Research* (New York: Time Books, Random House, 1995); see also Alan Stockdale, "Conflicting Perspectives: Coping with Cystic Fibrosis in the Age of Molecular Medicine" (Ph.D. diss., Brandeis University, 1997).

9. This summary is based on the work of Jean-Paul Gaudillière, "Biologie moléculaire et biologistes dans les années soixante: La naissance d'une discipline, Le cas français" (thèse pour le doctorat d'histoire des sciences de l'Université Paris VII, 1991); Moulin, *Le dernier langage de la médecine;* Anne-Marie Moulin and Ilana Lowy, "La double nature de l'immunologie: L'histoire de la transplantation rénale," *Fundamenta Scientiae* 4 (1983); Lowy, "Impact of Medical Practice on Biochemical Research."

10. Moulin, *Le dernier langage de la médecine,* p. 215.

11. Much can be said on the psychosocial roots of Daniel Cohen's comport-

ment and professional trajectory and on the relation of his charisma and ambivalence to his choices in science, in management style, and in his life.

12. Cohen, *Espoir,* p. 27.

13. Ibid., p. 44.

14. Ibid., p. 50.

15. Ibid., pp. 50–51.

16. Cook-Deegan, *Gene Wars,* p. 41.

17. Ibid., p. 42.

18. Ibid., p. 43.

19. Ibid., p. 44.

20. A popularized account of the emergence of the RFLP strategy is found in Jerry E. Bishop and Michael Waldholz, *Genome* (New York: Simon and Schuster, 1990).

21. Bernard Barataud, *Au nom de nos enfants* (Paris: Edition N° 1, 1992), p. 9.

22. Ibid., pp. 14–20.

23. Ibid., p. 224.

24. Ibid., pp. 233–35.

25. Ibid., p. 275.

26. Cohen, *Espoir,* p. 277.

27. On YACs, see Bertrand Jordan, "YAC Power," *BioEssays* 12, no. 4 (Apr. 1990); Hans M. Albertsen, Hadi Abderrahim, Howard Cann, Jean Dausset, Denis LePaslier, and Daniel Cohen, "Construction and Characterization of a Yeast Artificial Chromosome Library Containing Seven Haploid Human Genome Equivalents," *Proceedings of the National Academy of Sciences, U.S.A.,* 87 (June 1990); C. Bellanné-Chantelot, E. Barillot, B. Lacroix, D. LePaslier, and D. Cohen, "A Test Case for Physical Mapping of Human Genome by Repetitive Sequence Fingerprints: Construction of a Physical Map of a 42 kb YAC Subcloned into Cosmids," *Nucleic Acids Research* 19, no. 3 (1991).

28. Barataud, *Au nom,* p. 279.

29. An important demonstration of the power of YACs is found in Daniel Cohen et al., "Mapping the Whole Human Genome by Fingerprinting Yeast Artificial Chromosomes," *Cell* 70 (Sept. 18, 1992).

30. Barataud, *Au nom,* p. 281.

31. Cohen, *Espoir,* p. 19.

32. Ibid., p. 121.

33. Jordan, *Voyage autour du genome,* p. 130.

34. Ilya Chumakov, Daniel Cohen, et al., "Continuum of Overlapping Clones Spanning the Entire Human Chromosome 21q," *Nature* 359 (Oct. 1, 1992).

35. Cohen, *Espoir,* p. 196.

3

1. Cohen, *Espoir,* p. 163.

2. Ibid.

3. Ibid., p. 252.

4. Ibid., pp. 264, 266, 262.

5. Dominique Lecourt, *Contre la peur: De la science à l'éthique, une aventure infinie* (Paris: Hachette, 1990).

6. Cohen, *Espoir*, p. 225.

7. Ibid., p. 226. For what it is worth, I believe this talk is nonsense.

8. Ibid.

9. Ibid., p. 227.

10. Ibid., p. 233.

11. *CEPH-Info*, no. 3, p. 13.

12. Anne Carol, *Histoire de l'eugénisme en France* (Paris: Editions du Seuil, 1995).

13. Teams in Brazil and Japan began to work on similar studies at the same time that the CEPH did. See François Schachter, Laurence Faure-Delanef, Frédérique Guenot, Hervé Rouger, Philippe Froguel, Laurence Lesueur-Ginot, and Daniel Cohen, "Genetic Associations with Human Longevity at the APOE and ACE Loci," *Nature Genetics* 6 (Jan. 1994).

14. Lathrop has a massive bibliography. E.g., see G. M. Lathrop, J.-M. Lalouel, C. Julier, and J. Ott, "Multilocus Linkage Analysis in Humans: Detection of Linkage and Estimation of Recombination," *American Journal of Human Genetics* 37 (1985); G. M. Lathrop, D. Cherif, D. Julier, and M. James, "Gene Mapping," *Current Opinion in Biotechnology,* no. 1 (1990); G. M. Lathrop and J.-M. Lalouel, "Statistical Methods for Linkage Analysis," in *Handbook of Statistics,* vol. 8, ed. C. R. Rao and R. Chakraborty (Amsterdam: Elsevier, 1991).

15. Adam Telerman, Robert Amson, et al., "A Model for Tumor Suppression Using H-1 Parvovirus," *Proceedings of the National Academy of Sciences, U.S.A.,* 90 (Sept. 1993); Robert Amson, Adam Telerman, et al., "Isolation of 10 Differentially Expressed cDNAs in p53-Induced Apoptosis: Activation of the Vertebrate Homologue of the *Drosophila* Seven in Absentia Gene, *Proceedings of the National Academy of Sciences, U.S.A.,* 93 (Apr. 1996).

16. Peng Liang and Arthur B. Pardee, "Differential Display of Eukaryotic Messenger RNA by Means of the Polymerase Chain Reaction," *Science* 257 (Aug. 14, 1992).

17. Markoulatos Panaytotis, Maurice Moncany, et al., "HIV1 Related Sequences Detected by PCR in Seronegative Multitransfused Thalassemic Patients," *Journal of Viral Diseases* 1, no. 3 (1993).

4

1. Norbert Elias, *The History of Manners,* trans. Edmund Jephcott 1939; reprint (New York: Pantheon Books, 1978), originally published in 1939.

2. For a clear journalistic summary of these issues (that appeared as *French DNA* was in press), see Douglas Starr, *Blood: An Epic History of Medicine and Commerce* (New York: Alfred A. Knopf, 1998).

3. Grmek, *History of AIDS,* p. 161.

4. Ibid., p. 164.

5. Randy Shilts, *And the Band Played On: Politics, People, and the AIDS Epidemic* (New York: St. Martin's Press, 1987), p. 220.

6. Ibid., p. 224.

7. Ibid., p. 223.

8. Ibid., p. 477.

9. Ibid., p. 539.

10. Laurent Greilsamer, *Le procès du sang contaminé* (Paris: Editions Le Monde, 1993), p. 22.

11. Other countries were more prudent. Belgium avoided Factor VIII, employing a local alternative instead. Others adopted the heated standard more rapidly than France did. For details, see Starr, *Blood,* chap. 15, "Outbreak."

12. Created in 1949, amended under the laws of July 21, 1952, and *décret* of Jan. 16, 1954, and Sept. 7, 1958.

13. Quoted in Greilsamer, *Sang contaminé,* p. 26.

14. "Le risque d'être atteint du sida demeure pour un hémophile extrêmement faible.... Il n'y a aucune corrélation entre l'apparition du sida et la quantité des concentrés injectés" (quoted in Greilsamer, *Sang contaminé,* p. 40).

15. Grmek, *History of AIDS,* p. 165.

16. A highly placed French scientist working at a prestigious institution implicated in this affair verbally assaulted me at an international conference and threatened legal action from his institution if I claimed they had been in any way involved in endangering public health. I include the following quote to support my remarks. "The minutes from a May 9th meeting of the prime minister's cabinet spelled out their concern: 'The moment the tests are authorized the French market will be largely captured by the American test.... [Therefore,] the cabinet of the prime minister requests ... that the Abbott registration dossier be retained for some time by the National Public Health Laboratory.' Another confidential government memo stated that the objective should be to guarantee Pasteur about 35 percent of the national market. Meanwhile, Dr. Robert Netter, director of the National Public Health Laboratory, had proposed a strategy. 'Under the current circumstances it does not seem possible to delay registration any longer without risking a [charge of] abuse of power,' he wrote. 'I would therefore consider giving the Institut Pasteur an immediate registration and delaying Abbott's.' The Health Ministry acquiesced and approved Pasteur's AIDS test on June 21, 1985, a month before it granted Abbott's license. Pasteur would gain its foothold in France—at a price of several months' delay and thousands of untested donations" (Starr, *Blood,* p. 311). Starr adds (p. 407), "The sequence of events in the delay, along with exhibits, is detailed in M. Lucas, *Transfusion Sanguine et SIDA en 1985* (Paris: Inspection Générale des Affaires Sociales, 1991), pp. 45–51."

17. Dominique Memmi closes her book on the French National Ethics Committee by noting that they had a difficult task "in articulating, in a secular and scholarly language, the sacred" (*Les gardiens du corps: Dix ans de magistère bioéthique* [Paris: Editions de l'Ecole des Hautes Etudes en Sciences Sociales, 1996], p. 242).

18. Richard Titmuss in his classic *The Gift Relationship: From Human Blood to Social Policy,* ed. Ann Oakley and John Ashton, exp. and updated ed. (New York: New Press, 1997), notes that the British system was one of the very few that neither paid nor reimbursed, making it purer than the French system. Data on pp. 238–39.

19. Ibid., p. 314.

20. Ibid., p. 16. Furthermore, Vanessa Martiew pinpoints a decision taken in Britain in 1980 to include a "terminal heat treatment of coagulation factor

concentrates for 72 hours at 80C" as a measure against hepatitis. ("Transfusion Medicine towards the Millennium," in Titmuss, *The Gift Relationship,* p. 44). It is not exactly clear when these measures were put into effect, as the British center did not open until 1986.

21. Starr, *Blood,* p. 289.

22. "Mme Myriam Ezratty, aujourd'hui premier président de la cour d'appel de Paris, hier directrice de l'administration pénitentiaire, et signataire, en janvier 1984, d'une circulaire encourageant les collectes dans les prisons. . . . Me Nicole Dreyfus releva à quel point la contamination des lots de sang fut grande à l'occasion de ces prélèvements à la chaîne. . . . Elle s'indigna de ces tournées des centres de transfusion sanguine organisées dans des lieux où les donneurs à risque sont massivement représentés" (Greilsamer, *Sang contaminé,* p. 97).

23. Marie-Angéle Hermitte, *Le sang et le droit: Essai sur la transfusion sanguine* (Paris: Editions du Seuil, 1996), p. 186.

24. Ibid., p. 17.

25. Ibid., pp. 101, 93.

26. Ibid., pp. 111, 113.

27. Ibid., p. 102.

28. Ibid., pp. 106, 120.

29. Ibid., p. 73.

30. Ibid., p. 74.

31. Jean-Pierre Baud, *L'affaire de la main volée: Une histoire juridique du corps* (Paris: Editions du Seuil, 1993), p. 206.

32. Ibid., p. 207.

33. Hermitte, *Le sang,* p. 68.

34. Dominique Memmi, *Les gardiens du corps: Dix ans de magistère bio-éthique* (Paris: Editions de l'Ecole des Hautes Etudes en Sciences Sociales, 1996), p. 27.

35. Ibid., p. 15.

36. Jacques Testart, *L'oeuf transparent* (Paris: Flammarion, 1986), p. 29.

37. Ibid., p. 34.

38. For details, see Paul Rabinow, "Fragmentation and Redemption in Late Modernity," in *Essays on the Anthropology of Reason* (Princeton: Princeton University Press, 1996). There is a vast legal literature on the Moore case. Some of it is summarized and criticized by James Boyle in *Shamans, Software, and Spleens: Law and the Construction of the Information Society* (Cambridge: Harvard University Press, 1996), chap. 9, "Spleens."

39. Baud, *L'affaire,* p. 19.

40. Memmi, *Gardiens,* p. 25.

41. Claire Ambroselli, *Le Comité d'Ethique* (Paris: Presses Universitaire de France, collection "Que sais-je?" 1990).

42. Baud, *L'affaire,* p. 15.

43. David Rothman, *Strangers at the Bedside: A History of How Law and Bioethics Transformed Medical Decision Making* (New York: Basic Books, 1991), p. 63.

44. Ibid., p. 53.

45. Quoted in ibid., p. 176.

46. Renée C. Fox and Judith P. Swazey, "Medical Morality Is Not Bioethics: Medical Ethics in China and the United States," in *Essays in Medical Sociology,* ed. Renée C. Fox (New York: Wiley, 1979), pp. 668–70.

47. Immanuel Kant, *Foundation of the Fundamental Principles of the Metaphysics of Morals,* trans. Thomas K. Abbott (Indianapolis: Library of Liberal Arts, 1949), p. 51.

48. "L'être humain a droit à une protection absolue quand il s'agit du respect de la dignité de sa personne, la dignité inhérente à tous les membres de la famille humaine et de leurs droits égaux et inaliénables."

49. "Les mots 'libre,' 'dignité,' 'raison' n'ayant pas été, comme tels, l'objet de discussions durant les travaux préparatoires, il convient sans doute de les comprendre dans les limites fussent-elles large, il est vrai, du sens commun" (Albert Verdoodt, *Naissance et signification de la Déclaration Universelle des Droits de l'Homme,* Société d'Etudes Morales, Sociales, et Juridiques, Louvain [Louvain-Paris: Edition Nauwelaerts, n.d.], p. 84).

50. Memmi, *Gardiens,* p. 33.

51. Ibid., p. 15.

52. Ibid., p. 17.

53. Reminiscent of "touche pas à mon pote."

54. "Par respect pour son corps, dont il est responsable, mais non propriétaire, l'homme doit s'abstenir de le mutiler volontairement en dehors des cas où cet acte est nécessaire pour la santé de la personne" (ibid., p. 19).

55. Ibid., p. 18.

56. Cook-Deegan, *Gene Wars,* pp. 355–56.

57. Gerard Huber and François Gros, Avant-Propos to Gerard Huber, *Colloque: Patrimoine génétique et droits de l'humanité: Livre blanc des recommandations* (Paris: Osiris, 1990), p. 5.

58. Gerard Huber, "Rapport général," in *L'analyse du genome humain: Libertés et responsabilités,* Colloques, Associations Descartes, Dec. 7, 1992, part 3.4.2.

59. Elias, *History of Manners,* p. 50.

60. Ibid., p. 45.

61. Ibid.

62. Jacques Donzelot, *L'invention du social* (Paris: Editions Fayard, 1984).

63. Memmi, *Gardiens,* p. 242: "en énonçant, dans une langue laïque et savante, le sacré."

64. Memmi, ibid.: "dans les formes aujourd'hui socialement audibles."

65. Memmi, ibid.: "ce qu'il est digne ou indigne de faire au corps humain au nom des grandes causes médicale et scientifique."

66. Pierre Bourdieu, *Méditations pascaliennes* (Paris: Editions du Seuil, 1997), p. 288.

67. Anne Fagot-Largeault, "Respect du patrimoine génétique et respect de la personne," *Esprit* 5 (May 1991): 51.

5

1. Examples of Froguel's publications during this period are P. Froguel, D. Cohen, et al., "Close Linkage of Glucokinase Locus on Chromosome 7p to Early-Onset Non-Insulin Dependent Diabetes Mellitus," *Nature* 356, Letters section (Mar. 12, 1992); N. Vionnet, P. Froguel, D. Cohen, et al., "Nonsense Mutation

in the Glucokinase Gene Causes Early-Onset Non-Insulin Dependent Diabetes Mellitus," *Nature* 356, Letters section (Apr. 23, 1992).

2. André Chastel, "La notion de patrimoine," in *Les lieux de mémoire,* vol. 2, *La nation,* ed. Pierre Nora (Paris: Editions Gallimard, 1986), p. 405.

3. Ibid., p. 419.

4. Ibid., p. 405.

5. The blending of nature and culture into a socialized environment is explored in Paul Rabinow, *French Modern: Norms and Forms of the Social Environment* (Chicago: University of Chicago Press, 1996).

6. Martine Rèmond-Gouillard, "L'avenir du patrimoine," *Esprit* 216 (Nov. 1995): 66.

7. Jean Carbonnier, *Flexible droit,* 6th ed. (Paris: Librairie Générale de Jurisprudence, 1988), p. 256 n. 11.

8. "L'avenir," p. 67.

9. Ibid., p. 69.

10. Ibid., p. 72.

11. A short history of French representations of money and power is Pierre Birnbaum, *Le peuple et les gros: Histoire d'un mythe* (Paris: Grasset, 1979).

12. On "solidarity," see Rabinow, *French Modern,* esp. chap. 6.

13. *Impact-Médecin Hebdo,* no. 227 (Mar. 11, 1994): 28.

14. There is a huge literature on Vichy. On the Old Regime, see Michel Foucault and Arlette Farge, *Le désordre des familles: Lettres de cachet des Archives de la Bastille* (Paris: Editions Gallimard, Collection Archives, 1982).

15. "French Geneticists Split over Terms of Commercial Use of DNA Bank," *Nature* 356, no. 162 (1994): 175.

16. "French Gene Mappers at Crossroads," *Science,* Mar. 18, 1994, p. 1553.

17. Ibid.

6

1. An IPO is a public stock offering.

2. Gail Edmondson, "Come Home, Little Startups, Europe Wants Its Own NASDAQ to Capture More IPOs," *Business Week,* Feb. 26, 1996, p. 50.

3. *Nature* 379 (Feb. 8, 1996): 478.

4. Stephen Moore, "With Arrival of Research Star, Genset Embarks on Hunt," *Wall Street Journal,* Mar. 1, 1996, p. 5.

5. All quotations from Michael Balter, "Has French AIDS Research Stumbled?" *Science* 279 (Jan. 16, 1998): 312–14.

6. All quotations from "Tensions Grow over Access to DNA Bank," *Nature* 391, News section (Feb. 19, 1998).

7. All quotations from "What Ails French Biosciences?" *Science,* News and Comment, 279 (Mar. 6, 1998): 1442.

EPILOGUE

1. Cohen, *Espoir,* pp. 55–56.

2. George E. Marcus, "The Uses of Complicity in the Changing Mise-en-Scène of Anthropological Fieldwork," *Representations* 59 (summer 1997): 13.

3. Ibid., p. 16.

4. Foucault, *Surveiller et punir: Naissance de la prison* (Paris: Editions Galli-

mard, 1977), p. 35; *Discipline and Punish: The Birth of the Prison,* trans. Alan Sheridan (New York: Vintage Books, 1979), p. 31.

5. The expression "thick description" was introduced into the social sciences by Clifford Geertz, "Thick Description: Toward an Interpretive Theory of Culture," in *The Interpretation of Cultures* (New York: Basic Books, 1973). See also James Clifford and George Marcus, eds., *Writing Culture: The Poetics and Politics of Ethnography* (Berkeley and Los Angeles, University of California Press, 1986). The distinction between hermeneutics and interpretation is discussed in Hubert Dreyfus and Paul Rabinow, Introduction to *Michel Foucault: Beyond Structuralism and Hermeneutics,* 2d ed. (Chicago: University of Chicago Press, 1983). On interpretive social science, see Paul Rabinow and William Sullivan, eds., *Interpretive Social Science, A Reader* (Berkeley and Los Angeles: University of California Press, 1979); *Interpretive Social Science, A Second Look* (Berkeley and Los Angeles: University of California Press, 1987). A recent update of the entire trajectory of hermeneutics is Gerald Bruns, *Hermeneutics: Ancient and Modern* (New Haven: Yale University Press, 1992).

6. Paul Ricoeur, *Freud and Philosophy: An Essay on Interpretation* (New Haven: Yale University Press, 1970); Michel Foucault, "Nietzsche, Freud, Marx," in *Dits et écrits,* vol. 1, pp. 564–79 (originally published in *Cahiers de Royaumont,* vol. 6 [Paris: Editions du Minuit, 1967]).

7. John Rajchman, *Constructions* (Cambridge: MIT Press, 1998), p. 91.

8. Ibid., p. 116.

9. The concept of "background practices" is from Hubert Dreyfus, *Being-in-the-World: A Commentary on Heidegger's Being and Time, Div. I* (Cambridge: MIT Press, 1991).

10. "On croit volontiers qu'une culture s'attache plus à ses valeurs qu' à ses formes, que celles-ci, facilement, peuvent être modifiées, abandonnées, reprises: que seul le sens s'enracine profondément. C'est méconnaître combien les formes, où quand elles se défont qu'elles naissent, ont pu provoquer d'étonnement ou susciter de haine. Le combat des formes en Occident a été aussi acharné, sinon plus, que celui des idées ou des valeurs" (Michel Foucault, "Pierre Boulez: L'écran traversé," *Dits et écrits,* vol. 4, ed. Daniel Defert and François Ewald [Paris: Editions Gallimard, 1994], pp. 219–20; originally published in M. Colin, J. P. Leonardini, and J. Markovits, eds., *Dix ans et après: Album souvenir du festival d'automne* [Paris: Messidor, 1982], pp. 232–36).

11. Marshall Sahlins, *Island of History* (Chicago: University of Chicago Press, 1985), p. 153. The Sahlins quote is used by Liisa Malkki in her thoughtful attempt to grasp current events in refugee camps in Africa: "News and Culture: Transitory Phenomena and the Fieldwork Tradition," in *Anthropological Locations: Boundaries and Grounds of a Field Science,* ed. Akhil Gupta and James Ferguson (Berkeley and Los Angeles: University of California Press, 1997).

12. Thanks to Aihwa Ong for help on clarifying this passage and others.

13. Bruno Latour, *Science in Action: How to Follow Scientists and Engineers through Society* (Cambridge: Harvard University Press, 1987); Hans-Jorg Rheinberger, *Toward a History of Epistemic Things: Synthesizing Proteins in the Test Tube* (Palo Alto: Stanford University Press, 1997).

Bibliography

Agamben, Giorgio. *Homo sacer: Le pouvoir souverain et la vie nue*. Paris: Editions du Seuil, 1997. Translated by Daniel Heller-Roazen under the title *Homo sacer: Sovereign Power and Bare Life* (Stanford: Stanford University Press, 1998). Originally published as *Homo sacer. I.: Il potere sovrano e la nuda vita* (Turin: Giulio Einaudi, 1996).

Albertsen, Hans M.; Hadi Abderrahim; Howard Cann; Jean Dausset; Denis LePaslier; and Daniel Cohen. "Construction and Characterization of a Yeast Artificial Chromosome Library Containing Seven Haploid Human Genome Equivalents." *Proceedings of the National Academy of Sciences, U.S.A.*, 87 (June 1990).

Ambroselli, Claire. *Le Comité d'Ethique*. Paris: Presses Universitaire de France, collection "Que sais-je?" 1990.

Amson, Robert; Telerman, Adam; et al. "A Model for Tumor Suppression Using H-1 Parvovirus." *Proceedings of the National Academy of Sciences, U.S.A.*, 90 (Sept. 1993).

Arendt, Hannah. *The Human Condition*. Chicago: University of Chicago Press, 1958.

Balter, Michael. "Has French AIDS Research Stumbled?" *Science* 279 (Jan. 16, 1998).

Barataud, Bernard. *Au nom de nos enfants*. Paris: Edition No 1, 1992.

Barral, C.; and F. Paterson. "L'Association Française contre les Myopathies: Trajectoire d'une association d'usagers et construction associative d'une maladie." *Sciences Sociales et Santé* 12, no. 2 (1994).

Baud, Jean-Pierre. *L'affaire de la main volée: Une histoire juridique du corps*. Paris: Editions du Seuil, 1993.

Bellanné-Chantelot, C.; E. Barillot; B. Lacroix; D. LePaslier; and D. Cohen. "A Test Case for Physical Mapping of Human Genome by Repetitive Sequence Fingerprints: Construction of a Physical Map of a 42 kb YAC Subcloned into Cosmids." *Nucleic Acids Research* 19, no. 3 (1991).

Benison, Saul. "The History of Polio Research in the United States: Appraisal and Lessons." In *Twentieth Century Sciences: Studies in the Biography of Ideas,* ed. Gerald Holton. New York: W. W. Norton, 1972.

Berridge, Virginia. "AIDS and the Gift Relationship in the UK." In *The Gift Relationship: From Human Blood to Social Policy,* by Richard M. Titmuss, ed. Ann Oakley and John Ashton. Expanded and updated ed. New York: New Press, 1977.

Birnbaum, Pierre. *Le peuple et les gros: Histoire d'un mythe.* Paris: Grasset, 1979.

Bishop, Jerry E.; and Michael Waldholz. *Genome.* New York: Simon and Schuster, 1990.

Blumenberg, Hans. *The Legitimacy of the Modern Age.* Trans. Robert M. Wallace. Cambridge: MIT Press, 1983. Originally published in 1966.

Boltanski, Luc; and Laurent Thevenot. *De la justification: Les économies de la grandeur.* Paris: Editions Gallimard, 1991.

Bourdieu, Pierre. *Méditations pascaliennes.* Paris: Editions du Seuil, 1997.

Boyle, James. *Shamans, Software, and Spleens: Law and the Construction of the Information Society.* Cambridge: Harvard University Press, 1996.

Bruns, Gerald. *Hermeneutics: Ancient and Modern.* New Haven: Yale University Press, 1992.

Carol, Anne. *Histoire de l'eugénisme en France.* Paris: Editions du Seuil (1995).

Catala, Pierre. "La transformation du patrimoine dans le droit civil moderne." *Revue trimestrielle de droit civil,* no. 185 (1966).

Chastel, André. "La notion de patrimoine." In *Les lieux de mémoire,* vol. 2, *La nation,* ed. Pierre Nora. Paris: Editions Gallimard, 1986.

Chumakov, Ilya; Daniel Cohen; et al. "Continuum of Overlapping Clones Spanning the Entire Human Chromosome 21q." *Nature* 359 (Oct. 1, 1992).

Clifford, James; and George Marcus; eds. *Writing Culture: The Poetics and Politics of Ethnography.* Berkeley and Los Angeles: University of California Press, 1986.

Cohen, Daniel. *Les gènes de l'espoir: A la découverte du genome humain.* Paris: Robert Laffont, 1993.

Cohen, Daniel; et al. "Mapping the Whole Human Genome by Fingerprinting Yeast Artificial Chromosomes." *Cell* 70 (Sept. 18, 1992).

Cohen, D[aniel]; I. Chumakov; and J. Weissenbach. "A First-Generation Physical Map of the Human Genome." *Nature* 366 (Dec. 16, 1993).

Cook-Deegan, Robert. *The Gene Wars: Science, Politics, and the Human Genome.* New York: W. W. Norton, 1994.

Creager, Angela. "Campaigns and Crystals: The War against Polio and the Political Economy of Postwar Virus Research." In *The Life of a Virus: Wendell Stanley, TMV, and Experimental Systems as Models in Biomedical Research, 1930–1965.* Forthcoming.

Dausset, Jean. *Clin d'oeil à la vie: La grande aventure HLA.* Paris: Editions Odile Jacob, 1998.

Dausset, Jean; Howard Cann; Daniel Cohen; Mark Lathrop; Jean-Marc Lalouel;

and Ray White. "Program Description: Centre d'Etude du Polymorphisme Humain (CEPH): Collaborative Genetic Mapping of the Human Genome." *Genomics,* no. 6 (1990).

de Certeau, Michel. "Poétiques d'espace." *Café,* no. 1 (1983).

Donzelot, Jacques. *L'invention du social.* Paris: Editions Fayard, 1984.

Dorozynski, Alexander. "Gene Mapping the Industrial Way." *Science* 256 (April 24, 1992).

Dreyfus, Hubert. *Being-in-the-World: A Commentary on Heidegger's Being and Time, Div. I.* Cambridge: MIT Press, 1991.

Dreyfus, Hubert; and Paul Rabinow. Introduction to *Michel Foucault: Beyond Structuralism and Hermeneutics.* 2d ed. Chicago: University of Chicago Press, 1983.

Edmondson, Gail. "Come Home, Little Startups, Europe Wants Its Own NASDAQ to Capture More IPOs." *Business Week,* Feb. 26, 1996.

Elias, Norbert. *The History of Manners.* Trans. Edmund Jephcott. New York: Pantheon Books, 1978. Originally published in 1939.

Epstein, Steven. *Impure Science: AIDS, Activism, and the Politics of Knowledge.* Berkeley and Los Angeles: University of California Press, 1996.

Fagot-Largeault, Anne. "Respect du patrimoine génétique et respect de la personne." *Esprit* 5 (May 1991).

Feher, Ferenc; and Agnes Heller. *Biopolitics.* Public Policy and Social Welfare, vol. 15. Aldeshot: Avebury Publishers, 1994.

Foucault, Michel. *Discipline and Punish: The Birth of the Prison.* Trans. Alan Sheridan. New York: Vintage Books, 1979. Originally published as *Surveiller et punir: La naissance de la prison* (Paris: Editions Gallimard, 1977).

———. *History of Sexuality,* vol. 1. Trans. Robert Hurley. New York: Vintage Books, 1978. Originally published as *La volonté de savoir* (Paris: Editions Gallimard, 1976).

———. *The Order of Things: An Archaeology of the Human Sciences.* New York: Pantheon, 1970.

———. "Pierre Boulez: L'écran traversé." In *Dits et écrits,* vol. 4, ed. Daniel Defert and François Ewald. Paris: Editions Gallimard, 1994.

———. "What Is Enlightenment?" In *The Foucault Reader,* ed. Paul Rabinow. New York: Pantheon Books, 1984.

Foucault, Michel; and Arlette Farge. *Le désordre des familles: Lettres de cachet des Archives de la Bastille.* Paris: Editions Gallimard, Collection Archives, 1982.

Fox, Renée C.; and Judith P. Swazey. "Medical Morality Is Not Bioethics: Medical Ethics in China and the United States." In *Essays in Medical Sociology,* ed. Renée C. Fox. New York: Wiley, 1979.

"French Gene Mappers at Crossroads." *Science* 263 (Mar. 18, 1994).

"French Geneticists Split over Terms of Commercial Use of DNA Bank." *Nature* 356, no. 162 (1994).

Froguel, P.; D. Cohen; et al. "Close Linkage of Glucokinase Locus on Chromosome 7p to Early-Onset Non-Insulin Dependent Diabetes Mellitus." *Nature* 356, Letters section (Mar. 12, 1992).

Gaudillière, Jean-Paul. "Biologie moléculaire et biologistes dans les années soixante: La naissance d'une discipline, Le cas français." Thèse pour le doctorat d'histoire des sciences de l'Université Paris VII, 1991.

Geertz, Clifford. "Thick Description: Toward an Interpretive Theory of Culture." In *The Interpretation of Cultures*. New York: Basic Books, 1973.

Greilsamer, Laurent. *Le procès du sang contaminé*. Paris: Editions Le Monde, 1993.

Grmek, Mirko D. *History of AIDS: Emergence and Origin of a Modern Pandemic*. Trans. Russell Maulitz and Jacalyn Duffin. Princeton: Princeton University Press, 1990. Originally published as *Histoire du SIDA: Début et origine d'une pandémie actuelle* (Paris: Editions Payot, 1989).

Hermitte, Marie-Angéle. *Le sang et le droit: Essai sur la transfusion sanguine*. Paris: Editions du Seuil, 1996.

Hoffman, Eric P. "The Evolving Genome Project: Current and Future Impact." *American Journal of Human Genetics* 54 (1994).

Huber, Gérard. "Rapport général." In *L'analyse du genome humain: Libertés et responsabilités*. Colloques, Associations Descartes, Dec. 7, 1992.

Huber, Gérard; and François Gros. Avant-Propos to Gerard Huber, *Colloque: Patrimoine génétique et droits de l'humanité: Livre blanc des recommandations*. Paris: Osiris, 1990.

Jordan, Bertrand. *Voyage autour du genome: Le tour du monde en 80 labos*. Paris: Les Editions INSERM, 1993.

———. "YAC Power." *BioEssays* 12, no. 4 (Apr. 1990).

Kant, Immanuel. *Foundation of the Fundamental Principles of the Metaphysics of Morals*. Trans. Thomas K. Abbott. Indianapolis: Library of Liberal Arts, 1949. Originally published as *Grundlegung zur Metaphysic der Sitten* (1785).

Kuisel, Richard. *Seducing the French: The Dilemma of Americanization*. Berkeley and Los Angeles: University of California Press, 1993.

Lacorne, Denis; Jacques Rupnik; and Marie-France Toinet; eds. *The Rise and Fall of Anti-Americanism: A Century of French Perception*. Trans. Gerry Turner. New York: St. Martin's Press, 1990.

Lathrop, G. M.; Cherif, D.; Julier, D.; and James, M. "Gene Mapping." *Current Opinion in Biotechnology*, no. 1 (1990).

Lathrop, G. M.; and J.-M. Lalouel. "Statistical Methods for Linkage Analysis." In *Handbook of Statistics*, vol. 8, ed. C. R. Rao and R. Chakraborty. Amsterdam: Elsevier, 1991.

Lathrop, G. M.; J.-M. Lalouel; C. Julier; and J. Ott. "Multilocus Linkage Analysis in Humans: Detection of Linkage and Estimation of Recombination." *American Journal of Human Genetics* 37 (1985).

Latour, Bruno. *Science in Action: How to Follow Scientists and Engineers through Society*. Cambridge: Harvard University Press, 1987.

Lecourt, Dominique. *Contre la peur: De la science à l'éthique, une aventure infinie.* Paris: Hachette, 1990.

Le Goff, Jacques. *La bourse ou la vie.* Paris: Hachette, 1986. Translated by Patricia Ranum under the title *Your Money or Your Life: Economy and Religion in the Middle Ages* (New York: Zone Books, 1988).

———. *La naissance du purgatoire.* Paris: Editions Gallimard, 1981. Translated by Arthur Goldhammer under the title *The Birth of Purgatory* (Chicago: University of Chicago Press, 1984).

Liang, Peng; and Arthur B. Pardee. "Differential Display of Eukaryotic Messenger RNA by Means of the Polymerase Chain Reaction." *Science* 257 (Aug. 14, 1992).

Lowy, Ilana. "The Impact of Medical Practice on Biochemical Research: The Case of Human Leucocyte Antigens Studies." *Minerva* 25, nos. 1–2 (1987).

———. "Tissue Groups and Cadaver Kidney Sharing: Socio-cultural Aspects of a Medical Controversy." *Journal of Technology Assessment in Health Care* 2 (1986).

Malkki, Liisa. "News and Culture: Transitory Phenomena and the Fieldwork Tradition." In *Anthropological Locations: Boundaries and Grounds of a Field Science,* ed. Akhil Gupta and James Ferguson. Berkeley and Los Angeles: University of California Press, 1997.

Marcus, George E. "The Uses of Complicity in the Changing Mise-en-Scène of Anthropological Fieldwork." *Representations* 59 (summer 1997).

Martiew, Vanessa. "Transfusion Medicine towards the Millennium." *Impact-Médecin Hebdo,* no. 227 (Mar. 11, 1994).

Mathy, Jean-Philippe. *Extrême-Occident: French Intellectuals and America.* Chicago: University of Chicago Press, 1993.

Memmi, Dominique. *Les gardiens du corps: Dix ans de magistère bioéthique.* Paris: Editions de l'Ecole des Hautes Etudes en Sciences Sociales, 1996.

Moore, Stephen. "With Arrival of Research Star, Genset Embarks on Hunt." *Wall Street Journal,* Mar. 1, 1996.

Moulin, Anne-Marie. *Le dernier langage de la médecine: Histoire de l'immunologie de Pasteur au SIDA.* Paris: Presses Universitaire de France, 1991.

Moulin, Anne-Marie; and Ilana Lowy. "La double nature de l'immunologie: L'histoire de la transplantation rénale." *Fundamenta Scientiae* 4 (1983).

Mounier, Emmanuel. "De la propriété capitaliste à la propriété humaine" (1936). In *Flexible droit: Pour une sociologie du droit sans rigueur,* ed. Jean Carbonnier. 6th ed. Paris: Librairie Générale de Jurisprudence, 1988.

Nature 370, no. 6484 (July 7, 1994).

Nature 379, no. 6565 (Feb. 8, 1996): 478.

Nelson, Benjamin. *The Idea of Usury: From Tribal Brotherhood to Universal Otherhood.* 2d ed. Chicago: University of Chicago Press, 1969.

Nietzsche, Frederic. "Mortal Souls." In *Daybreak: Thoughts on the Prejudices of Morality.* Trans. R. J. Hollingdale. Cambridge: Cambridge University Press, 1982. Originally published in 1886.

Panaytotis, Markoulatos; Moncany, Maurice; et al. "HIV1 Related Sequences Detected by PCR in Seronegative Multitransfused Thalassemic Patients." *Journal of Viral Diseases* 1, no. 3 (1993).

Patterson, James T. *The Dread Disease: Cancer and Modern American Culture.* Cambridge: Harvard University Press, 1987.

Pinell, Patrice. *Naissance d'un fléau: Histoire de la lutte contre le cancer en France, 1890–1940.* Paris: Meraillé, 1992.

Pollak, Michel. *Les homosexuels et le SIDA: Sociologie d'une épidémie.* Paris: A. M. Metailie, 1988.

Raberharisoa, Vololona; and Callon, Michel. "L'implication des malades dans les activités de recherche soutenues par l'Association Française contre les Myopathies." *Sciences Sociales et Santé* 16, no. 3 (1998).

Rabinow, Paul. "Fragmentation and Redemption in Late Modernity." In *Essays on the Anthropology of Reason.* Princeton: Princeton University Press, 1996.

———. *French Modern: Norms and Forms of the Social Environment.* Chicago: University of Chicago Press, 1996. Originally published in 1989.

———. *Making PCR: A Story of Biotechnology.* Chicago: University of Chicago Press, 1996.

Rabinow, Paul; and William Sullivan; eds. *Interpretive Social Science, A Reader.* Berkeley and Los Angeles: University of California Press, 1979.

———. *Interpretive Social Science, A Second Look.* Berkeley and Los Angeles: University of California Press, 1987.

Rèmond-Gouilloud, Martine. "L'avenir du patrimoine." *Esprit* 216 (Nov. 1995).

Rajchman, John. *Constructions.* Cambridge: MIT Press, 1998.

Rheinberger, Hans-Jorg. *Toward a History of Epistemic Things: Synthesizing Proteins in the Test Tube.* Palo Alto: Stanford University Press, 1997.

Ricoeur, Paul. *Freud and Philosophy: An Essay on Interpretation.* New Haven: Yale University Press, 1970.

Rimbaud, Artur. Letter to Paul Demeny, May 15, 1871. In *Poésies,* by Artur Rimbaud. Paris: Bookking, 1991.

Rothman, David. *Strangers at the Bedside: A History of How Law and Bioethics Transformed Medical Decision Making.* New York: Basic Books, 1991.

Sahlins, Marshall. *Island of History.* Chicago: University of Chicago Press, 1985.

Schachter, François; Laurence Faure-Delanef; Frederique Guenot; Herve Rouger; Philippe Froguel; Laurence Lesueur-Ginot; and Daniel Cohen. "Genetic Associations with Human Longevity at the APOE and ACE Loci." *Nature Genetics* 6 (Jan. 1994).

Scott, Joan Wallach. *Only Paradoxes to Offer: French Feminists and the Rights of Man.* Cambridge: Harvard University Press, 1996.

Sève, Lucien. *Pour une critique de la bioéthique.* Paris: Editions Odile Jacob, 1994.

Shilts, Randy. *And the Band Played On: Politics, People and the AIDS Epidemic.* New York: St. Martin's Press, 1987.

Smith, Jane S. *Patenting the Sun: Polio and the Salk Vaccine.* New York: Anchor Books, 1990.

Starr, Douglas. *Blood: An Epic History of Medicine and Commerce.* New York: Alfred A. Knopf, 1998.

Stockdale, Alan. "Conflicting Perspectives: Coping with Cystic Fibrosis in the Age of Molecular Medicine." Ph.D. diss., Brandeis University, 1997.

Telerman, Adam; Robert Amson; et al. "Isolation of 10 Differentially Expressed cDNAs in p53-Induced Apoptosis: Activation of the Vertebrate Homologue of the *Drosophila* Seven in Absentia Gene." *Proceedings of the National Academy of Sciences, U.S.A.,* 93 (Apr. 1996).

"Tensions Grow over Access to DNA Bank." *Nature* 391, News section (Feb. 19, 1998).

Testart, Jacques. *L'oeuf transparent.* Paris: Flammarion, 1986.

Titmuss, Richard M. *The Gift Relationship: From Human Blood to Social Policy.* Ed. Ann Oakley and John Ashton. Expanded and updated ed. New York: New Press, 1997. Originally published in 1970.

Verdoodt, Albert. *Naissance et signification de la Déclaration Universelle des Droits de l'Homme.* Société d'Etudes Morales, Sociales et Juridiques, Louvain. Louvain-Paris: Edition Nauwelaerts, n.d.

Vionnet, N.; P. Froguel; D. Cohen; et al. "Nonsense Mutation in the Glucokinase Gene Causes Early-Onset Non-Insulin Dependent Diabetes Mellitus." *Nature* 356, Letters section (Apr. 23, 1992).

Vovelle, Michel. *Les âmes au purgatoire, ou le travail du deuil.* Paris: Editions Gallimard, 1996.

Weber, Max. *The Protestant Ethic and the Spirit of Capitalism.* Trans. Talcott Parsons. New York: Charles Scribner's Sons, 1958. Originally published in 1904–5.

———. "Religious Rejections of the World and Their Directions" and "Science as a Vocation." In *From Max Weber: Essays in Sociology,* ed. H. H. Gerth and C. Wright Mills. New York: Oxford University Press, 1946.

Wexler, Alice. *Mapping Fate: A Memoir of Family, Risk, and Genetic Research.* New York: Time Books, Random House, 1995.

"What Ails French Biosciences?" *Science* 279, News and Comments (Mar. 6, 1998).

Acknowledgments

My thanks go first to present and former Berkeley students, friends of wisdom all: James Faubion, Joao Biehl, Steven Collier, and Andrew Lakoff. Second, I offer my gratitude to all those in France—especially at the Centre d'Etude du Polymorphisme Humain—who generously provided their *sagesse,* expertise, time, and habitus. To protect the innocent and to avoid providing fuel for innuendo, I do not specify names. Third, a group without whose support of various kinds this work would not have taken the shape it did deserves mention: Suzanne Calpestri, Lawrence Cohen, Angela Creager, Mike Fortun, Duana Fulwilley, Christian Girard, Douglas Holmes, Robert Hurley, George Marcus, Jonathan Marks, Donald Moore, Aihwa Ong, Adriana Petryna, Francis Pisani, and Tom White. Fourth, I thank Marilyn Seid-Rabinow and Marc Rabinow for their energy and endurance, as well as the wonders of life as I know it. Fifth, I offer official and sincere acknowledgment for the support of the National Science Foundation (Professional Development Fellowship, 1994), the material and spiritual aid of the French INSERM Unité 158, and, finally, to the University of California's President's Fellowship in the Humanities.